INTRUSIVE LEADERSHIP

HOW TO BECOME THAT LEADER
2nd Edition

MARCUS A. CANADY

Intrusive Leadership

INTRUSIVE LEADERSHIP

HOW TO BECOME THAT LEADER
2nd Edition

MARCUS A. CANADY

Intrusive Leadership, Becoming THAT Leader, *2nd Edition*
ISBN 978-1-966480-00-6 Softbound
ISBN 978-1-966480-08-2 Hardbound
ISBN 978-1-966480-09-9 EBook
Copyright © 2024 Marcus A. Canady

Request for information should be addressed to: Curry Brothers Marketing and Publishing Group
P.O. Box 247 Haymarket, VA 20168

Cover Design by Vibranium Media
Manuscript Editing by J&S Company (Chantel Johnson & Tiffany Johnson)

CURRY BROS.
MARKETING + PUBLISHING GROUP

TABLE OF CONTENT

Daring to "Intrude"

When Marcus Canady first approached me about writing the foreword for this book on "intrusive leadership," I'll admit I had a visceral reaction to the term. In my decades of work as a DEI consultant and advocate for inclusive leadership, the word "intrusive" seemed to run counter to everything we strive for in creating psychologically safe workplaces where people can bring their full selves. Yet as I dove deeper into Marcus's work, I realized that this apparent contradiction holds a profound truth about what effective leadership requires today.

The work of evolving as leaders, and our vocabulary to describe that evolution, needs a big shake-up and update in these polarizing times. It's of paramount importance that we continue to be creative and search for provocative concepts and terminology to spark us and others to grow and develop new skills, and I credit Marcus with provoking a change in everyone who encounters this concept and wrestles with the key question I ask myself, every day: "Am I doing enough of the right things to create a more inclusive future?"

Every few years, a leadership style or principle emerges as the new hot topic in corporate and academic circles. Whether it's emotional intelligence, servant leadership, authentic leadership, or change management, these concepts capture our collective imagination and quickly become the buzzwords on every leadership expert's tongue. My own career as a DEI consultant came in on the crest of the inclusive leadership wave, and I'm still riding it today because its relevance and necessity have only grown stronger.

The very notion that leadership should be "intrusive" might make us uncomfortable – and that's precisely the point. True leadership, like meaningful allyship and advocacy, often requires us to push beyond our comfort zones, to intentionally disrupt the status quo, and yes, sometimes to "intrude" upon the familiar patterns that keep exclusion and inequity in place.

What I find particularly compelling about Marcus's framework is how it aligns with the developmental journey I've observed in inclusive leaders. In my work with thousands of leaders through my Inclusive Leader Continuum, I've seen how the capacity for meaningful "intrusion" evolves across distinct phases of leadership growth.

It begins with interrupting ourselves – our own patterns, assumptions, and biases. This initial intrusion is internal, as we move from being unaware to questioning our default ways of thinking and leading. We must first "intrude" upon our own comfort zones before we can effectively engage with others. This often involves confronting uncomfortable truths about our blind spots and privileges.

As leaders progress in their journey, intrusion takes the form of intentionally placing ourselves in new conversations, networks, and learning spaces. We begin to "intrude" upon our familiar circles, seeking out diverse perspectives and experiences that challenge our understanding. This awareness phase requires us to acknowledge what we don't know and actively work to expand our viewpoint.

The next evolution involves taking active steps to change how we lead. Intrusive leadership at this stage means deliberately interrupting "business as usual" – speaking up in meetings when we notice exclusion, changing how we make decisions to include more voices, and challenging our teams to think differently about inclusion. We begin to "intrude" upon established norms and practices that may be limiting our organization's potential.

Finally, at the advocate level, intrusive leadership becomes bold and systemic. Leaders at this stage regularly "intrude" upon power structures and decision-making processes that perpetuate inequity. They ask uncomfortable questions in board rooms, challenge succession planning norms, and actively work to reshape organizational systems. Their intrusion becomes a powerful force for institutional change.

This evolution in leadership practice comes at a crucial time. The speed of technological change is faster than ever. The global competition for quality talent is more intense than ever. The stressors on people's lives are arguably more demanding now than ever before. As a result, the way we lead must evolve to meet these challenges.

The emerging workforce–diverse, vocal, and increasingly unwilling to compartmentalize their identities – demands more from leadership. They seek leaders who will actively engage with their whole selves, who will "intrude" upon the traditional boundaries between professional and personal, not to violate them, but to honor the full humanity of each person. This younger generation doesn't want to code-switch or present a different persona at work. They want to feel passionate about their work while having their employers demonstrate genuine care for them as complete human beings.

The stories and examples Marcus shares throughout this book illustrate what I've seen in countless organizations: that transformative leadership happens when we dare to move beyond transactional relationships to create authentic human connections. Whether it's a leader reaching out to check on an employee during times of social upheaval, or a supervisor taking the time to understand the full context of someone's performance challenges, these "intrusions" become the building blocks of trust, belonging, and ultimately, organizational success.

In an era where mental health challenges are prevalent, where global events impact our workplace experiences in real-time, and where younger generations expect their leaders to engage with social issues, intrusive leadership provides a framework for meeting these complex demands. Leaders who will excel in the future will be those who are intentional about earning trust, willing to engage in difficult conversations, and committed to personalizing their leadership approach to meet individual needs.

The leadership revolution Marcus describes isn't just about changing how we lead – it's about transforming how we show up for each other as human beings in the workplace. In a world where many feel increasingly disconnected despite our digital connectivity, intrusive leadership offers a path toward creating the authentic connections and belonging that we all seek.

As you read this book, I encourage you to lean into any discomfort the term "intrusive leadership" might provoke. Consider where you are in your own journey as an inclusive leader, and how you might need to "intrude" differently – whether that's beginning to question your own assumptions, actively seeking new perspectives, challenging established norms, or advocating for systemic change. Often, it's in these moments of discomfort that our greatest opportunities for growth and transformation emerge.

Jennifer Brown

Best-Selling Author, *How to be an Inclusive Leader: Your Role in Creating Cultures of Belonging where Everyone can Thrive; Podcast Host, The Will to Change*

Endorsements

"The second version of this masterpiece is a welcome companion and needed as a follow-up to "Intrusive Leadership: How to be THAT Leader." Whether you have one direct report or thousands that you lead, people are our most important resource and sure the mission is very important but will fail if you don't take care of the people charged with executing the mission. Trust me, this book is a must read and necessary for your leadership library. Marcus has outdone himself again!"

Director Kevin M. Sligh
Bureau of Safety and Environmental Enforcement

Leaders want to optimize the performance of the team. In Intrusive Leadership, Marcus Canady provides leaders the tools they need to understand their most important resource: people. Unless this is the very first book you've read about leadership, you will quickly recognize concepts like trust, empathy, transparency, mentoring and listening. Canady ties these and other leadership concepts together with the call for leaders to take action and become actively engaged in understanding the men and women they work with. Intrusive Leadership is full of examples that are very relatable and make Canady's theme come alive.

Rear Admiral Andrew Lennon
U.S. Navy (retired)

Intrusive Leadership" is an insightful and transformative read for anyone in a leadership role. Marcus Canady masterfully combines personal anecdotes with actionable strategies, making complex leadership concepts both relatable and easy to implement. His emphasis on proactive and engaged leadership is a refreshing departure from traditional management literature and is desperately needed in today's divisive social and business climate.

Mr. Rich Brady
CEO, Society of Defense Financial Management

To be the best leader I can be, I seek models from others and Marcus Canady's "Intrusive Leadership" did not disappoint. Through his stories Marcus calls on every leader to dial it up and truly be there for those around them. Consider the power of these words when dealing with female US Coast Guard Academy cadets, regarding the lack of female Commanding Officers. Marcus notes "...there isn't one woman on that wall. Don't let that discourage you. Don't let that give you the idea you can't be on that wall....Let the fact that you don't see a woman motivate you to become the first one....take a selfie next to that wall. When things get tough on your journey, pull that picture out and remind yourself of the goal...Now go make it happen." Wow! Marcus' words drive positive action.

Ms. Christine Landoll
Director of Business Engagement/Professor of Practice
George Mason University

In his book, "Intrusive Leadership, How to Become THAT Leader", Marcus Canady has clearly defined using anecdotal stories, the benefits of leaders who implement the intrusive leadership style. On the surface, using the word intrusive sounds like a leader should go to places in their subordinates' lives where they should not go. However, Marcus Canady explains that people bring their entire selves to work which includes any positive or negative situations they may be experiencing. Administering to the entire person instead of just the parts, enhances employee performance as well as the cohesion needed for effective teams and teamwork. This is a book for any leader who desires to take themselves and the people they lead to the next level.

Mr. Kenneth Hunter
Diversity/Equal Employment Opportunity Manager
DC Fire and EMS

Marcus is vulnerable and real. He opens his heart to explain this leadership style and models what it means to be an intrusive leader. He challenges you to become more self-aware, step outside your comfort zone, and strive to be a better leader. This book is not only a guide but an invitation to transform your approach to leadership and make a lasting impact on those you lead.

Ms. Patti Tutalo
CEO, Patti Tutalo Consulting

Intrusive leadership involves actively engaging with team members, which helps leaders understand their subordinates' strengths, weaknesses, aspirations and challenges. This deeper connection fosters trust and loyalty. Intrusive leaders are often more attuned to the issues and challenges their team members face. This proactive approach allows for early identification and resolution of problems, preventing small issues from escalating.

CAPT Janet Days
U.S. Navy (retired)

"This is the book every leader needs. If you want to be an effective leader, you have to know your people and what matters in their lives. Intrusive leadership is the key to successful teams. Marcus Canady provides powerful insight into why IL matters and why it works. With this book, you can build a more capable and winning team."

Cathy Barrington, Colonel, U.S. Air Force

"Marcus Canady uncovers timeless wisdom that arises from his experience of addressing real issues that aren't obvious to most. Intrusive Leadership isn't a slogan, it's a mindset for how the people we admire would encourage us to effectively lead others today, no matter our industry."

Eric P. Kowack, Senior Vice President for Xencor Corporation

"This is an artfully written book that is required reading for all leaders within my organization. Marcus Canady defines and discusses a leadership style that is impactful beyond measure. He challenges us to be better as leaders and create relationships with everyone in an authentic manner. This tool is invaluable to us as military leaders and applicable to all that seek the benefits of leading a cohesive team."

Alex "Waldo" Hampton, CAPT
US Navy Commander, Carrier Air Wing 7

"With Intrusive Leadership, Marcus Canady issues an impassioned call for leaders to pierce the layers of superficiality that are common in the boss-worker relationship, and instead connect on a meaningful level. Canady talked to renowned world leaders and everyday managers about their best and worst experiences. The theme he found was that interactions that tap into the humanity of workers are rare, appreciated, and often the basis of job satisfaction and retention. Using real world examples, Canady proves that the power and benefits of Intrusive Leadership are available to everyone, and that sensing how and when to apply it results in grateful, loyal, and motivated workforces. He calls on us to push past doubts and create workspaces where people feel seen, connected, and valued."

Dr. Terri A. Dickerson
U.S. Coast Guard Senior Executive and Adjunct Professor/George Mason University

"Captain Canady reminds us that leaders develop through practice. Practice becomes habit, and constant habit lets us achieve excellence. His book provides concrete examples of how to guide subordinates to their own individual success. A gem for anyone who chooses to be a leader."

Michelle J. Howard, Admiral, U.S. Navy, (Retired)

"In his book, Marcus Canady skillfully lays out a rational and compelling argument as to why leaders should adopt a style of intrusive leadership. Using sound research and rich storytelling, he advances the intrusive leadership concept from a being ill-defined and controversial to being specifically defined and potentially universal. The author's infusion of his own experiences adds to the captivating nature of the book.

As one who has stepped through all phases of the leadership continuum, from apprentice to senior executive, I rate the book as a must-have in any leadership library. I will add it to my ever growing leadership library as a loaner to mentees.

I recommend this book to anyone engaged in leading self, leading others, leading performance and change, or leading organization: From parents, teachers, coaches, and students from high school up… To public servants, military professionals and elected officials… To corporate and non-profit supervisors, executives and board members… And to leaders in fields such as education, healthcare, STEM, and sports and entertainment.

I know from personal experience that an intrusive leadership style, even under the most challenging of circumstances, can achieve optimum results. Thanks to the author's work, intrusive leadership has the potential to become universal and enduring. We will all be the better for that!"

Manson K. Brown, Vice Admiral, U.S. Coast Guard (Retired)
Former Assistant Secretary of Commerce

"Excellent treatise, I could not put it down!

Timely, relevant, poignant. Intrusive Leadership captures, clearly and succinctly, the critical elements of successful leadership in today's hyper-engaged world.

Intrusive Leadership offers a unique treatise to address today's leadership challenges. It proposes a proven, effective, interactive method for engaging anyone, and it generates an extraordinarily positive impact on their lives and careers. Thought-provoking and evidence-based, the book is replete with sterling, real life examples of its impact and importance.

The author crafts a clear and compelling case that Intrusive Leadership is a leadership style that uses intentional actions in a caring and engaging manner. Intrusive Leadership is a personal leadership style that is best achieved by drawing on the personal strengths of the leader and the needs of the employee.

Intrusive Leadership is a critical must-have in today's leadership tool kit."

Erroll Brown, Rear Admiral, U.S. Coast Guard (Retired)

"Insightful, inspiring and informative. Captain Marcus Canady's analysis and application on Intrusive Leadership provides a fresh perspective on what leadership needs to look like today and going into the future. His examples are compelling as he provides personal stories throughout his study to back up his hypotheses. There are far many other books on leadership, but this one is uniquely purposeful and useful. Government, corporate and academic leaders will find great value in reading this treatise."

Sinclair Harris, Rear Admiral, U.S. Navy (Retired)

"A rich and diverse workforce brings new dimensions of capability to any organization, but also new dimensions of responsibility for leadership. Marcus Canady provides a manual on how to meet the moment."

Joanna M. Nunan, Rear Admiral, U.S. Coast Guard (Retired)

"Over the years I have made it a professional quest to study leadership, which is the art of influencing behavior to accomplish a desired goal. Marcus A. Canady leverages the tenets of Intrusive Leadership by going beyond the standard merits of basic leadership by pivoting to a new way of approaching leadership with a "whole-body" experience. You simply can't lead by asking, or directing, so when using Intrusive Leadership, you must actively participate with the endeavor, by first engaging, and immersing yourself into getting to know all participates, so you can accurately gauge the outcome and results. Intrusive Leadership creates a new level of understanding, I mean truly understanding the strengths and weaknesses of every single member on your team. I seriously encourage you to listen closely to Canady's thesis, and embrace his teachings, because Intrusive Leadership is the future of enterprise leading in a civilized society."

Dr. Gerald D. Curry Senior Executive Service (SES), U.S. Air Force Director, Air Force Review Boards Agency

"Do you care enough to intrude when it's needed? Would you intrude into a house on fire to ensure everyone was safely out? Would you intrude in someone's life when they're hurting? Before reading this book I shared camp with a couple of folks you'll see interviewed in these pages who thought "intrusive" just wasn't the right word choice. CAPT Canady has convinced me otherwise. During our shared time at the National War College, we were privileged to call "Theodore Roosevelt Hall" on Ft. McNair our office space. TR is often credited with having first penned the sentiment, "Nobody cares how much you know until they know how much you care." This is Marcus' central message and leaders at every level should read this book!"

Jeff Hurlbert, Brigadier General, U.S. Air Force 32nd Commandant, National War College

"The subject on 'Intrusive Leadership'has not had as much of an intellectual discussion as it should have over the past decade, in my opinion. Intrusive Leadership can be viewed as a comprehensive, and sometimes a complex style of leadership, among the categories of effective ways to lead people. The author's explanation breaks it all down in more than just common terms but does an exceptional job of categorizing the characteristics of individuals in supervisory roles and responsibilities and how intrusive leadership provides the best solution as a principal tool in looking after the needs of those who are led.

We as leaders often fall into the delicate situation where the importance of ensuring that our subordinates, colleagues, and others we encounter are doing okay. There's a misconception that too much prying into a person's business is a step over the line, rather than the assurance and confidence that a problem or situation has been identified effectively. This is why I feel that the author's detailed explanation of intrusive leadership is so important and articulated well in this book.

The key element that this book gets across is WHY it's important to 'drill down' in detailed understanding of an individual within the workforce that anyone that holds the mantle of being 'in charge' in some manner, must take full advantage of utilizing his or her leadership styles to ensure that all is well within the environment. The examples that the author provided in this book should be viewed with interest and allow for the reader to take the opportunity to imagine within their own environments, what is going on, and possible options on how to respond to difficult situations when they arise, as well as anticipate.

The book, "Intrusive Leadership," while excellent for anyone in a managerial role to read and understand, is paramount for those who are directly involved in supervisory leadership roles. While it's excellent for reading and understanding in the military environment, it also is equally important for just about anyone employed in a private or government sector position, whether it's a front-line supervisor at a manufacturing plant, retail, restaurant, as well as office and technology settings. You don't need to be in a higher-up positional power role to grasp the concepts of this book's intention."

"In the science of leadership, we are guided by professional and personal core value statements that lay out our conditions of employment or action to focusing and carrying out tasks and responsibilities. Adding the contents of further defined explanation of intrusive leadership as outlined in this book, provides an exceptional guide to help become a better and informed leader."

Vince Patton, Ed.D.
Master Chief Petty Officer of the U.S. Coast Guard, (Retired)

"What makes an organization of people go the extra-mile? What makes a good leader GREAT? Marcus Canady has the answers in Intrusive Leadership. He doesn't just tell you how to execute Intrusive Leadership. He tells you the when, where, and why behind it. If you value mental health and the well-being of yourself and those you serve, read this book. You will walk away with impactful techniques you can incorporate today. You will walk away with an understanding that to lead is less about what you know and more about how well you care for those under your charge. Marcus Canady draws upon current social and racial tensions to weave a narrative that is both insightful and powerful. This is a must-read for every single junior officer in our military!"

Ashley S. Lee, Colonel, U.S. Army

Acknowledgments

I cannot begin this book without thanking a number of people that have impacted, motivated, and supported me along this journey. There is absolutely no way I would be here today, doing what I am doing, living out my dreams, if it wasn't for God's favor and the people he has placed in my life.

Thank you to the numerous intrusive leaders that have invested in me throughout my career. I have been blessed to work for some amazing people that took me under their wing and showed unwavering care and support for me as a person. From my time as a cadet at the US Coast Guard Academy, people have always encouraged me to keep going. Coaches, instructors, alumni, classmates, and especially the cafeteria workers went above and beyond to make sure that I knew someone was behind me cheering me on. Your words, your investments, the extra orange juice cartons at my lunchroom table made a difference. Thank you!

Thank you to my National War College (Class of 2022, Committee 10, Cherry Parrots, President's Cup winners) classmates and staff. I arrived at Roosevelt Hall during the summer of 2021 at an interesting point in my life. I believe I was mentally and emotionally exhausted. I needed the laughs, I needed the softball games, I needed the insightful conversations, and I needed the crawfish boil. You saw me frustrated, you saw me concerned, you saw me questioning, and you didn't look away. You engaged, you listened, and you took action. You provided me an opportunity to research Intrusive Leadership holistically in an academic fashion. You provided me the opportunity to engage with some senior members in the US government and that was absolutely life changing. This book would not have happened if it wasn't for this experience with you. Thank you!

To General Colin Powell. When I decided to research Intrusive Leadership, you were at the very top of the list of people I wanted to interview. When I think of one of the leading voices in this space, yours quickly pushes to the forefront. I never got the chance since you succumbed to COVID right as I began my journey, however I felt your presence. Spent hours in the library dedicated to you at the war college and even sat at your old desk. Thank you for the impact you had on me from afar, and for the impact you had across the world. You are a leadership example for everyone.

To my US Coast Guard shipmates. I have a lot more runway behind me than in front of me. I have been wearing the uniform for over 25 years at this point. I would do it all over again in a heartbeat. I would go through all the uniform inspections, days away from homeport, hours spent on general mandated training, and more. I would do it all knowing that I get to come to work every day with phenomenal people that want to make a difference. It has been an absolute honor to serve alongside you. I will forever be grateful. Thank you!

To my village. You know who you are, and you know how you have motivated me. When I am in meetings or at conferences, you are with me. You have provided the courage I needed to speak my truth to power because that truth is ours. You have recognized the battles because you knew I was fighting for you. You have challenged me. You have pushed me. You have held me accountable. You told me to give you a reason to come to graduation day. You made sure you were visible at my promotion ceremonies. You were there to check up on me. Throughout the years you have cheered the loudest and been the sharpest critics. You made me better. Iron sharpens iron. Thank you!

To my family. There is nothing in this world like love and support from those you call family. No matter where I have been or what I have done. The family support has been consistently tremendous. I have had cousins that traveled to see me graduate and take command. Aunts and uncles that have cheered me on. My parents and sister have spent time editing my work and helped sharpen my skills. My parents laid a firm foundation for me to build on. You were my first intrusive leaders. They planted a seed that they have watered every day of my life. I hope you are proud of the tree that has grown. Thank you family! Thank you Sabrina! And thank you Mommy and Daddy!

To my children, Marlisse, Major, and Maverick. Being your father is an absolute joy. On good days or bad days, coming home to you always brings a smile to my face. You three crumbsnatchers are amazing. You keep me going…even when you should be in the bed sleep. You keep me laughing. You provided me a new purpose. I am so thankful that God chose me to be your father. Thank you!

To my wife, my soulmate, my Proverbs 31, my best friend, my Angelisse. I can write an entire book on the joy you have brought to my life. I won't even begin to attempt to fully describe how valuable you are to Team Canady. You are phenomenal. I look forward to enjoying this journey with you and building up your pedestal along the way. To forever we go!!!

Why A 2nd Edition?

At the time that I am writing this 2nd Edition, it has been less than a year since the first edition was published and available for purchase. That isn't necessarily a long time to pass for the need to update the original version. Some authors wait 5, 10, or more years before they begin drafting their 2nd edition. So why am I doing it so soon? Was the first version wrong or incorrect, and I needed to quickly correct it? Nope, that isn't the reason. However, there is a reason that I am not waiting years to provide an update to this book.

The reason that I am not waiting to put out a 2nd edition goes back to my decision not to wait to put out the first edition. While I was researching Intrusive Leadership and conducting interviews with various people in corporate America, government agencies, and military branches, I made a conscious decision. I knew that more research needed to be done, more interviews, and more stories collected. I knew that there was more that needed to be said about Intrusive Leadership, but I didn't want to wait. I didn't want to wait two or three years to allow for more extensive research and produce a 400-page textbook. I didn't want to wait because I felt there was a need for Intrusive Leadership in workplaces at that very moment. I wanted to impact the way that people are being led immediately because of the feedback that I was receiving after every presentation or interview.

I have been traveling the country doing presentations and seminars on Intrusive Leadership since the Fall of 2022. Routinely after every presentation, I would get approached by someone who referenced how much they enjoyed what they heard and wished they had heard it years before. I had one military member say to me, "I am retiring in a month after serving 20 years, and this is the first presentation on leadership that I connected with." I heard story after story about how the lack of Intrusive Leadership was the main contributing factor for someone deciding to leave an organization or place of employment. I also heard heart-wrenching stories about leaders missing opportunities to support people and having them end up committing suicide. I just did not want to wait. I wanted the conversation about Intrusive Leadership to start immediately so I decided to publish the first edition in August of 2023. And I am so glad that I did.

Now it is July of 2024, and I have conducted over a hundred presentations on Intrusive Leadership from coast to coast. I have done multiple podcasts and engaged with thousands of people. My conviction that Intrusive Leadership will be the most critical leadership style for the future has only grown and strengthened. The future success of organizations or corporations' hinges on their ability to create an organizational culture where Intrusive Leadership is engrained in the organizational fiber. Where it is the common standard versus the highlighted exception. Supervisors must be held accountable for leadership skills and not just managerial skills. A new way of engaging employees must be trained and instituted to meet the needs and desires of the generational talent entering the workforce today.

In addition to my conviction growing stronger, my knowledge and perspective on Intrusive Leadership has grown and expanded as well. The first edition was nowhere near a finished product that included every aspect or detail of the leadership style. It included what I had witnessed, discovered, and learned about at the time. Some time has passed and now I have witnessed, discovered, and learned more. My leadership journey began a long time ago, and I intend for it to continue for a long time into the future. My goal is to take you on a journey, and I am so thankful that you have decided to join. If you read the first edition, then I deeply appreciate you staying with me, and I hope you find the additional stories and reflections meaningful. I have added some more characteristics of an intrusive leader and a section that addresses frequently asked questions that I encountered after my speaking engagements. I genuinely believe you will like how the conversation has evolved. If you have not read the first edition don't worry. Nothing has been deleted or changed, things have only been added and sections expanded. You hopped on the journey right when you were supposed to. Your timing is perfect. Let's begin…

About the Book

How do you become the leader THAT truly inspires people? How do you become the leader THAT people trust? How do you become the leader THAT saves someone's life? You become THAT leader by being an intrusive leader.

Intrusive Leadership: How to become THAT Leader is the first and only book solely dedicated to defining, dissecting, and discussing the most critical leadership style for the future. This book will take you on a journey from introducing the leadership style, identifying current problems, to eloquently stating the case that every person in a leadership role should know about Intrusive Leadership. Through the art of storytelling, you will be exposed to the undeniable impacts of when a person is led in a manner that makes them feel seen and valued.

The world is rapidly changing, and global economic competition is fierce. The companies and organizations that take the time to refine and adjust their leadership framework to match the needs of the younger generation will attract and retain the best talent. Their employees will be more loyal, more committed to organizational success, and more willing to make the necessary effort to realize their full potential. That leadership framework has to include Intrusive Leadership.

Let's start the journey together...

PART I

Introduction

Right off the bat I am going to tell you exactly what this book is about and the message I intend to deliver. If you are in the bookstore or online wondering if you should add this book to your cart, then hopefully this introduction will answer that question. No smoke and mirrors, no creative advertising to entice you, no bells and whistles. Just the simple and honest purpose of this book. I believe in the message enough that I think it doesn't need any of that other stuff.

The message that I intend to convey in this book is that Intrusive Leadership is a leadership style that every supervisor, manager, and executive in any organization or company needs to incorporate. Every teacher, coach, or police chief needs to understand the positive impacts of Intrusive Leadership. Anyone that is in or aspires to be in a leadership position and wants to have a positive impact over people's careers and lives needs to know how to use Intrusive Leadership. Friends and co-workers need to understand how Intrusive Leadership can save someone's life. It is a difference maker, and the necessity of it for successful companies and organizations will be more and more critical in the future. The people occupying leadership positions that choose not to adopt Intrusive Leadership will struggle to attract, motivate, and retain quality talent and thus will lose whatever strategic competition they are in.

In this book, I will explain how I became a champion for Intrusive Leadership. I will define it, dissect it, and provide examples of it. I will also explain how impactful this leadership style can be. I hope that by the end of this book you will agree on how important Intrusive Leadership is and will begin to incorporate it in your own leadership toolbag. If you are in a position of organizational influence, I hope to compel you to consider how to incorporate Intrusive Leadership into the culture and climate of your organization. If I fail in this goal and you disagree, then let's talk about it. I would love to hear your perspective and thoughts. Trust me, in the world of social media, you will find me.

People are without a doubt and without question the most important asset to any organization. How you value, nurture, and support that asset thus becomes the most important decision for any leader to make. Intrusive Leadership is a leadership style that places people at the center of focus. It is completely about their desires, their needs, their fears, their motivations, and more. It is about leading the whole person, not just the fraction of the person that you see during their work shift. It is a style that builds deep and trusting relationships that unlocks untapped potential and creates a work environment that positively feeds an employee's mind and soul.

My Journey

The Article

It was a sunny Friday morning in June of 2020. I was sitting in my truck, parked in my parking spot, and taking a moment to myself before I walked into my office. I needed a moment because I was struggling. I was struggling trying to process what I saw while watching the video of George Floyd's murder by a police officer in broad daylight, on an American street, in front of a crowd of people. I was already trying to process what happened to Ahmaud Arbery, Breonna Taylor, and what Christian Cooper had to endure just simply bird watching in a New York City public park. When I reflected on those names and situations, I couldn't help but think about Walter Scott, Sandra Bland, Freddie Gray, Tamir Rice, Trayvon Martin, Laquan McDonald, Botham Jean, Atatiana Jefferson, Jordan Edwards, John Crawford III. So many names…too many names.

As I thought about the names of African Americans recently killed by police officers, in police custody, or by someone trying to be a police officer…my mind kept going back to what happened to Christian Cooper. Here he was simply bird watching and only asked a middle-aged woman to put her dog on a leash as the dog was coming close to him. Dogs were required to be on a leash as per park policy. That simple request by Christian Cooper to Amy Cooper, a white woman of no relation, resulted in a frantic call to police and a false accusation of threatening violence.

Thankfully, Christian recorded the incident with his cellphone because odds are his honest words would not have been enough to defend himself. In the video, you can clearly see a white woman who understood her privilege and used it against an African American man that she knew was innocent. She knew that she could use the police as her weapon and that Christian was powerless, only being armed with the truth. You can clearly see her threaten him, call the police, and provide a false report of violence in a frantic voice. How many African American men are incarcerated today because of a frantic voice giving a false report? How many have been lynched? I thought about Emmett Till.

I was not okay. I needed to talk to someone. I needed to express my anger and frustration. I needed someone to hear my pain. I just needed someone to listen. But I was the Commanding Officer of Coast Guard Air Station Houston, Texas and I had a job to do. I had to support and serve the men and women doing phenomenal work to carry out lifesaving missions and law enforcement operations along the Texas gulf coast. I had to get myself together to be there for them. So I put on my mask, as so many African Americans have done well before the COVID-19 pandemic, and put on a show. A show to display everything was okay with me and there was a sincere smile behind the KN95 mask. I bottled up my emotions and went to work. I walked around the hangar deck and workspaces, greeting people like I normally did. Asking them how they were doing, how were things at home, making sure everyone else was okay. I finally made it to my office where I could safely stop the show, take off the mask, and be my authentic self….alone.

I remained alone and secluded until a junior officer showed up at my office to deliver me a gift. He was stationed at another nearby Coast Guard unit, and he was another African American. When we locked eyes, not a word was said but a lot was understood. He was not okay either. COVID-19 social distancing protocols were tossed to the side. We dapped each other up, gave a much-needed brotherly embrace, and we talked. He was going through the same emotions I was after seeing the same video that the world saw of George Floyd being murdered.

His emotions were amplified because there was actually a discussion about the situation at his unit. Some of his co-workers expressed outrage and frustration as they watched the news. They wanted something to be done, something to change. They were upset about the broken glass windows of businesses and burning buildings in Minneapolis. They were emotionally impacted by the property damage and not the loss of George Floyd's life. What they wanted changed was people's reaction to the injustice, not the injustice itself. Nobody asked this junior officer how he was doing or about his feelings. Nobody saw this as a moment where he needed to be cared for and supported so that he could be the best version of himself at work. He needed that and I needed that.

That evening when I got home, I started thinking that there had to be more people in need as well. There had to be more African Americans that were finding it difficult to compartmentalize their emotions and focus on their jobs. In many cases, struggling to focus while doing dangerous jobs that require undivided attention. I wanted to do something to let supervisors and Commanding Officers know that this was a moment that couldn't be overlooked. This was an opportunity to connect with a group of people in a way that hasn't been done before. To show this group that they were valued and that their life mattered. I began to write and what I wrote ended up becoming the online article "Racial Tension in America Requires Military Intrusive Leadership"1 for the U.S. Naval Institute's Proceedings journal. This is what I wrote:

Even as racial tensions flare across the United States, I can be present at quarters, lead a preflight brief, pass critical information to the unit or even hand out awards as the commanding officer. Yet, there is a part of me that is invisible at work—because I wear a mask that hides the part that is hurting, tired, and frustrated. As an African American, I wear this mask because it is at times what I have to do to be included and accepted in the workplace. It is what I do in the hope of being treated like my fellow shipmates. I feel that it is expected—if not required—for me to have a chance at achieving my professional goals and safely executing my missions in the Coast Guard. It is what I have to do to continue to provide for my family.

What is my mask hiding? It hides the confident and proud African American side of me from which I draw strength and motivation. However, this side of me needs support at times. Even this source of strength can be depleted. Currently my mask hides the side of me that is struggling to process the recent killings of Ahmaud Arbery, Breonna Taylor, and George Floyd — in the nation I love and serve in the year 2020. The mask hides the side of me that is hurting as if I had lost a family member. It hides a side of me angered by these injustices. It hides the side of me that is not okay at this moment.

When the Coast Guard faced its lapsed appropriation during the 2019 government shutdown, senior leaders communicated to commanding officers across the service how important intrusive leadership was in that moment. It was understood that the financial strain on our members might have an emotional impact that could not be overlooked. Supervisors were directed to go the extra mile to make sure the members of their commands

---

1. "Racial Tension in America Requires Intrusive Military Leadership," U.S. Naval Institute, June 2, 2020, https://www.usni.org/magazines/proceedings/2020/june/racial-tension-america-requires- intrusive-military-leadership.

knew they were valued and that leaders understood the struggle members were going through.

In other difficult times, too, intrusive leadership has been seen as necessary to address increases in suicide, depression, and domestic violence. Commanding officers and supervisors are frequently instructed to look for signs that indicate there might be something going on outside the workplace that is affecting members of their commands deeply. Intrusive leadership currently is being invoked to help servicemembers deal with the COVID-19 pandemic. The fact that members' families and social lives have been impacted and the norms of life and work have been disrupted, once again, has supervisors asking, "Are you okay?"

The emotional impact, the struggle for social justice, and the disruption to normal life from the tragic events involving African Americans are worthy of that same level of intrusive leadership. The masks that many African American Coast Guard members have been wearing since they first reported to officer training in New London or boot camp at Cape May are not on so tight that emotions are invisible. Some perceptive shipmates are good at seeing those emotions. Others are less perceptive or perhaps choose to ignore the emotions of their African American shipmates because they feel that only other African Americans can share the pain and frustration.

There is a need for intrusive leadership in this moment. Many of us are hurting and finding it difficult to get dressed for upcoming law enforcement boardings or to relieve the watch on the quarterdeck. Some of us cannot escape when liberty is piped because our homes are on cutters or in the barracks. Some of us are feeling a shortness of breath because our masks have started to choke us. On the mess decks or in the coffee mess, when conversations about violence arise only after the looting starts, and not when another innocent heartbeat is stopped, we struggle to carry out our daily military tasks. There is a need for intrusive leaders who see that we are hurting right now. We need caring, empathetic leaders who can look behind our masks and say that even though they don't walk in the same shoes we do, they know that during these turbulent times the shoes we wear are a little heavier today.

I did not know Ahmaud Arbery, Breonna Taylor, or George Floyd, but I grieve the loss of their lives. Any time a life is taken, and it is believed that the color of the victim's skin is part of the reason, it hits deep. People with that same skin color, including me, are affected personally as if the victim were one of our own family members. We might not need to take emergency leave, but we still are working through the stages of grief. Simply being asked if we are okay during times like these can lift us up and help us breathe.

As a commanding officer, executive officer, or command master chief, you might learn that some members at your unit do not feel included. You might learn that recreational areas in berthing spaces are not a safe place for some of your crew to decompress. You might learn something about your crewmembers that will allow you to lead and motivate them better. You might even learn something about yourself.

I invite supervisors and senior leaders across the Coast Guard (and the entire military) right now to ask a simple question of the African Americans in your commands: "Are you okay?" You don't have to understand all the emotions, all the factors, or have similar opinions to show that you care. You did not have all the answers when you asked the same question during the lapse of appropriations last year, but you still asked. Some members were financially stable because they had savings or other sources of income, but you still asked. You asked because you cared enough to make sure that everyone was okay, and you were willing to take the risk of being an intrusive leader with some who did not need it. Not every African American feels the way I do. Some are okay right now; others are struggling deeply, and their masks are hiding a lot. Whether they are feeling okay or struggling, none will mind the question. All will appreciate leaders who care enough to ask. And those caring, intrusive leaders will help make us all better prepared to safely and professionally execute our missions.

The Article Aftermath

Immediately after the article hit the Proceedings website and was shared on social media, my phone and emails were flooded with messages of support, appreciation, and gratitude. Friends, mentors, academy classmates, and my supervisors reached out. All of them thanking me for opening their eyes to a new and different perspective. Some apologized for failing to reach out before being prompted by the article. I received numerous emails and social media comments from people I never met, telling me that I wrote the words that they held inside and afraid to let out.

Years later I still vividly remember some of the email messages and phone conversations. One email message from a Senior Enlisted Petty Officer expressed her ability to take a deep breath for the first time since watching the George Floyd video. One phone conversation with a white male Chief Petty Officer who was stationed with a young African American male in Georgia, close to where Ahmaud Arbery was murdered. This Chief Petty Officer expressed his difficulties with this crime. He wanted to reach out and talk to this young man and he wanted my advice on how to approach the difficult but necessary conversation. The article was also shared throughout the Coast Guard by senior officers and people began to talk to each other in a way they have never talked to each other before. I was invited to speak to a few units and participated in a couple of panel discussions focused on racism and inclusion. Overall, I feel that the article had a positive impact and people understood the necessity of Intrusive Leadership in that moment.

There are a few stories that resulted from the article that I would like to expand on. The first story is about the change in work relationship between CDR Ryan Miller and Master Chief Petty Officer Cliff Cook. Both of them worked at the same Coast Guard unit in Galveston, Texas and had a fairly simple and normal working relationship. Focused on the job, doing the mission, and rarely taking the time to get to know each other. That changed after CDR Miller read the article. These two co-workers started to talk about things that were going on outside of work. They started to talk about where they grew up, how they were raised, the differences in their interactions with law enforcement.

CDR Miller grew up in a small town in western Pennsylvania, where interactions with minorities were few and far between. Master Chief Cook grew up in inner city Mobile, Alabama where the rare interactions with white people were often police officers and unpleasant.

They talked about their families and the conversations Master Chief had to have with his son that CDR Miller never even thought of having with his kids. They began to have a relationship that went deeper than the daily worklist. They got to know each other and learn from each other. Through personal and intrusive interactions, their normal work relationship got better.

Now I have known Master Chief Cook for over 25 years, and I am a better man because of it. He is one of my closest friends. I would argue that anyone would be better after getting to know him. It would have been an utter shame if CDR Miller had spent two years working side by side with Master Chief Cook and never received the benefit of truly getting to know him. I'm glad my article made a difference for the both of them.

The second story is about Steve and Celia Peace and their relationship with their granddaughters. This lovely couple attends the same church as my parents in Hill Country just outside San Antonio, Texas and received a copy of my article from my father. They read my article that was focused on improving work relationships and workplace climates in military organizations and wondered if it could impact their family relationships.

This white couple had a daughter who married an African American man and their marriage gave them two wonderful granddaughters. Two wonderful bi-racial granddaughters. The ethnicity of their son-in-love (I stole this term from my mother-in-love who told me that it is love that brought us together, not the law) never was a factor, issue, or concern for Steve and Celia Peace. They have seen bigotry and despised it. They didn't raise their daughter to harbor bigoted views.

They judged everyone on their character as Martin Luther King Jr. had directed. The only thing that mattered was how their son-in-love treated and cared for their daughter. She was happy therefore they were happy parents. They were also happy and loving grandparents. They had a great relationship with their granddaughters and enjoyed spending time with them. I'm sure they spoiled them in a way only grandparents can do. Ethnicity was never an issue for this family, but it was also never a topic of conversation. The experiences their granddaughters were having growing up as bi-racial children and the things they were exposed and subjected to wasn't discussed with their grandparents. That changed after reading my article. After reading my article they decided to discuss how it was being an African American with my parents. One evening at my parents' home, they also discussed events and experiences growing up in Texas, traveling the country as a military family, and raising my sister and me in predominantly white suburban neighborhoods. The conversation was impactful and the Peace's eyes were opened to a reality of a different life in America. They discovered that even though they had worshipped on Sunday mornings at the same church and been in small group bible study classes, their lives were truly worlds apart. From this interaction with my parents, they decided to talk to their son-in-love and their granddaughters.

They asked their son-in-love about his upbringing and they learned about *"The Talk."* Not the talk about the birds and the bees, the talk about how to survive a traffic stop by the police. A talk that has taken place for generations in African American households across the country. A talk that was absolutely foreign to Steve Peace. They asked their granddaughters how they were dealing with what was happening in America in the Spring of 2020 and they heard stories of what their granddaughters had been dealing with for years. They heard about making their white friends mad when they played around with their African American friends and being made fun of by their African American friends for not being "black" enough. They heard stories about how they were treated and how they witnessed the treatment of their father. How they were angry and afraid at times, but how they had each other to lean on.

The Peaces were shocked, they were surprised, and their love and support for their granddaughters got even stronger. Stronger because it touched an area of their lives that it didn't touch before. You can be truly supportive and caring without being intrusive. You can have a good relationship and enjoy the company of others without being intrusive. But when you are intrusive, it enhances that care and support. It turns a good relationship into a great one and it makes the joyful company of others impactful. The Peaces have a better understanding of what their granddaughters are experiencing in their lives and now they can support and care for them in a more holistic manner. These conversations provided Steve and Celia with a new perspective and a keener awareness to things that were happening around them. They became better friends to my parents, better in-loves to their son-in-love, and better grandparents. They can take the spoiling of the granddaughters to a whole new level.

The third and final story I want to share is what happened to me at my unit. At the time that I wrote the article, I was the only African American uniformed member at Coast Guard Air Station Houston. There was one civilian employee but due to COVID and opposing work schedules we hadn't seen each other in a few weeks. I worked alongside some of the best men and women that this great country has to offer. People that were experts in their job and selfless in their service. I worked alongside awesomeness everyday… but at that moment I felt alone and isolated. I felt that since I was the only African American uniformed member then that meant that I was the only one emotionally impacted by what was playing out on city streets. I thought I was the only one upset that a police officer could calmly press the life out of another American citizen in handcuffs in broad daylight. I thought I was the only one questioning how much progress we have actually made in regards to race relations. I was wrong, I wasn't alone.

After my article started to circulate online and was shared by senior Coast Guard officers, member after member approached me about their own anger. They discussed their own disgust at what they saw. They were emotionally impacted just like I was and I had stereotyped them. I needed to talk to someone about my feelings and so did they. I had an opportunity to be the Commanding Officer they needed in that moment and I had missed it. They saw me as any one of the hundreds of Commanding Officers in Coast Guard history and they also recognized me as just the fourth African American Commanding Officer of an Air Station in Coast Guard history. They made me better and I will always be grateful that they did.

After I witnessed the reaction of people who read the article, heard the stories of the impact that it had within the Coast Guard and beyond, and had my own leadership ability improved…a fire was lit. After reflecting on the type of leaders that have impacted me in my career, conducting countless interviews, and researching leadership styles and principles, that fire has only grown. There is a void in the realm of leadership, and I am aiming to fill it. This is just the beginning.

Well…that was the beginning and nothing has changed from how this journey started. What began the passion and fire, or the aim of filling that leadership void hasn't changed either. I will share with you one observation that I have encountered more than a few times when I share my story.

On a few occasions I have had some attendees get up and walk out of the presentation or seminar the moment I mention George Floyd's name. They don't cause a disturbance or announce why they are leaving. However, their abrupt departure upon hearing George Floyd's name was noticeable. I have had comments made to me by other attendees indicating that they noticed the silent protest as well. Now, keep in mind that during my presentations I don't explain my opinion or thoughts surrounding the actions or factors that led to George Floyd's murder. I just simply say that watching the video of the murder and reflecting on other situations and racial tension at that moment was having an impact on my emotions.

If someone simply stating how something impacted them causes a person to no longer want to engage in a leadership discussion, then that is a person who is never going to be an intrusive leader. My guess, is that those individuals made a snap decision about my thoughts and assumed that it didn't align with their thoughts or opinions on the matter. They felt so strongly about their perspective being the right perspective that they had no desire to learn about another. That is a clear sign that those individuals do not want to be forced to meet people where they are, making an effort to connect with certain people, and try to become a better leader for every person they supervise. My guess, is that those people will choose to have "their favorites" and only build relationships with people who have a similar mindset. While only engaging with others solely as employees versus as human beings.

These are exactly the types of people who need to be identified, provided Intrusive Leadership training that they can't walk out of, held accountable for displaying the actions of an intrusive leader, and potentially let go if they simply don't adjust. Company leadership needs to realize they have these types of people within their ranks, and they need to decide what they are going to do with them. I intend to make those people uncomfortable during my presentations because I suspect those same people are making other people feel uncomfortable in workspaces every single day.

Intrusive Leadership is not for individuals who want to remain comfortable or individuals who want to avoid sensitive and difficult conversations. Management is often comfortable, but leadership will often put you in uncomfortable situations (more on that later). After one presentation I had a white man approach me so he could share his thoughts after listening to my message. He was currently in a mid-level management position, and he explained that he has always tried to build personal relationships with his employees. However, he intentionally avoids discussing anything that deals with ethnicity, race, or discrimination. He thought that it was best to just not go there and make sure every interaction and discussion was a happy and comfortable one. He now realized that he might have had good intentions but potentially ended up communicating that he didn't care about that aspect of his employees. He realized that he constrained how deep the relationships he was building was going to go and in-turn limited his impact as a leader.

PART II

Current Challenges with Intrusive Leadership

The term Intrusive Leadership is not a new term. I can't claim to be the person to invent the term. It has been used by senior leaders in the military in isolated situations for decades. While researching Intrusive Leadership for an academic paper while attending the National War College (Class of 2022, Committee 10, Cherry Parrots, President's Cup Champions), I came across a few articles.

In 2004, Rear Admiral Dick Brooks, US Navy (retired), wrote the article "Committed to Protecting Our People" in the Naval Safety Center's aviation magazine called Approach. In this article he stated, "An overall, across the board reduction of mishaps requires intrusive leadership and everyone's dedicated efforts."[2] In 2007, Vice Admiral Jeffrey Fowler, US Navy (retired), wrote " 'Develop' is an action verb requiring intrusive leadership from staff, faculty, and coaches." when discussing how to develop US Naval Academy Midshipmen into future military leaders.[3] Admiral Mike Mullen, while serving as the Chairman of the Joint Chief of Staff, delivered remarks about shifting some focus to the psychological fitness of soldiers at the Association of the U.S. Army luncheon in 2010. He stated, "This profound operational shift will also require the renewal in engaged, focused, and in some cases, very intrusive leadership."[4] Three different senior military leaders telling people and directing their subordinates to exercise Intrusive Leadership to increase operational safety, develop young military men and women, and to positively impact the mental health of soldiers.

---

2. Dick Brooks, "Committed to Protecting Our People," Approach 49, no. 4 (July 1, 2004)
3. VADM Jeffrey Fowler, "Naval Academy . . . A Crucible for Warriors," U.S. Naval Institute, October 1, 2007, https://www.usni.org/magazines/proceedings/2007/october/naval-academy- crucible-warriors.
4. "Remarks by Admiral Michael Mullen, USN, Chairman of the Joint Chiefs of Staff (Cjcs), at the Association of the U.s. Army (Ausa) Sustaining Member Luncheon Location: Washington, D.c. Date: Wednesday, October 27, 2010," Congressional Hearing Transcript Database, October 27, 2010, https://nduezproxy.idm.oclc.org/login?url=https://search.ebscohost.com/login.aspx?direct=tru e&AuthType=ip,url,uid&db=edsgao&AN=edsgcl.24 0922191&site=eds-live&scope=site.

In more recent times, the final investigation report into the unfortunate deaths of a couple of US Army soldiers at Fort Hood (now Fort Cavazos) listed some findings and recommendations to avoid these situations in the future. The first finding listed in the report, discussed the Sexual Harassment Assault Response and Prevention (SHARP) program at Ft. Hood. The report identified that things were in place for the SHARP program to be in compliance. However, there was insufficient leadership on a consistent basis to truly impact the culture and climate in the desired manner. The report stated, "If ever there was a need for intrusive hands-on leadership with regards to the health and welfare of troops, Fort Hood is and was the environment."[5] The report goes on to list several recommendations. The fifth recommendation under the section to address Command Climate issues was to practice Intrusive Leadership in these areas through direct and persistent engagement.

Suicide is a major issue that is plaguing the military services and civilian organizations right now. The number of people turning to self-harming actions because they are unable to cope with depression, anxiety, and stress is astonishing. Senior military leaders are dedicated to finding ways to support their members and reverse this trend. A 2021 memorandum in the U.S. Navy directed commanders to use Intrusive Leadership as one of the measures to address this suicide problem.[6] An Associated Press article in December of 2022 discussed the investigative report following three suicides within one week on one Navy Aircraft carrier. The report stated that there wasn't a link between the suicides, but it also stated that this was a time for intrusive leadership.[7]

I can personally recall the term Intrusive Leadership being used during my military career in two distinct situations: the U.S. Coast Guard lapse of appropriation and the recent COVID-19 pandemic. Some military officials used the term, while others directed supervisors to carry out the actions of

---

5. "2020-12-03_FHIRC_report_redacted.Pdf," accessed February 25, 2022, https://www.army.mil/ e2/downloads/rv7/forthoodreview/2020-12-03_FHIRC_report_redacted.pdf.
6. Christen McCurdy, "Navy Directs Commanders to Follow up after Suicide Prevention Referrals - UPI.Com," United Press International, January 28, 2021, https://www.upi.com/ Defense-News/2021/01/28/Navy-directs-commanders-to-follow-up-after-suicide-preven-tion- referrals/3581611879693/.
7. "Navy Finds No Link in Series of Suicides by Carrier Crew," accessed January 24, 2023, https://www.usnews.com/news/politics/articles/2022-12-19/navy-finds-no-link-in-series-of-suicides-by-carrier-crew.

Intrusive Leadership. In 2018, during the winter holiday season, a budget dispute in Congress led to a historic government shutdown. Active-duty military men and women in the U.S. Coast Guard worked and stood the watch without getting paid.

At the time I was serving as the Aviation Operations Officer at Sector/Air Station Corpus Christi, Texas. I was responsible for all the training and mission execution for the pilots and aviation crews flying three HC-144 airplanes and three MH-65D helicopters. I loved this job for many reasons, but during this situation I loved it for a completely different reason. I love it because I was able to watch America's best sons and daughters show up every day, do their jobs, and stand the watch for the American people knowing that they were not getting paid. Selfless service to the maximum.

These remarkable actions didn't cover up the stress and agony of financial uncertainty however. Some of our members faced the reality of not being able to pay bills or buy groceries. Military members don't plan on not getting paid while serving their country. A job still had to be done and done safely. For the first time that I can remember, I was directed to check on my people and ask specifically how they were managing things at home. I directed others to do the same. There was a clear recognition that people could be dealing with situations at home that they could not easily compartmentalize. What was happening outside of work could impact their performance at work. We couldn't ignore the operational safety issue the lapse in appropriation was creating; Intrusive Leadership was our solution.

We started directing aviation crews to discuss how they were specifically doing financially before every flight. We consistently reminded people that unit leaders were there to help them manage the problems they were facing at home. We told them to trust us enough to come and talk to us. Not every member needed help, but everyone appreciated the willingness and desire their supervisors displayed in a way that they couldn't have imagined. Intrusive Leadership helped my unit come together during a difficult time and continue to safely execute the mission.

In 2020, we faced another situation where Intrusive Leadership was thrust to the forefront of leadership conversations. the global COVID-19 pandemic. Billions of people across the world had their normal daily routines disrupted by a health crisis that took the lives of over a million people. People that normally went to work and interacted with other individuals, were forced to stay at home and interact through computer screens. Work conferences were canceled, the sports world was put on pause, routine date nights ceased. There was no March Madness for the first time in my life. Extroverts were not doing okay.

At Coast Guard Air Station Houston, we couldn't do our jobs from home, so we did come to work. We shifted to a one and four rotation and eventually settled into a one and three rotation. I went to work once every three days with the same group of people. Days of the week only mattered to keep up with the TV show schedules. The daily routine of life for everyone was impacted.

Tragically, a lot of lives were taken from our society. People were dealing with not having daily interactions with loved ones compounded with losing loved ones. Losing loved ones and then not even able to come together as a family, celebrate life, and comfort one another during the difficult time. A lot of people were not doing okay with the daily routine disruptions.

Suicides throughout society increased. This additional stress was recognized by my senior leaders in the Coast Guard and by me as well. The District Commander in Coast Guard District Eight (senior executive position in the civilian world) started to schedule routine phone calls with his unit Commanding Officers. The focus of these phone calls was not for him to get updated on mission operations or for him to provide task direction. The focus of these phone calls was solely about making sure we, his Commanding Officers in the field, were doing okay mentally and emotionally with the COVID-19 pandemic.

During my phone calls with Rear Admiral John Nadeau, U.S. Coast Guard (ret), we talked about my family a lot. I told him about the routine walks around the neighborhood with my wife, three kids, mother-in-love, and Johnny Bacon (the family dog). I told him how excited my two older kids were when we went to get Dairy Queen ice cream and they got to enjoy a car ride for the first time in months. I didn't realize how much my 4-year old daughter and 2-year old son were also impacted until I saw the joy on their faces for simply being out of the neighborhood. Rear Admiral Nadeau and I also talked about what was happening in our society with racial tension at a high level. We talked about personal experiences growing up and how the Coast Guard was addressing diversity, equity, and inclusion. A phone meeting on my work schedule to solely discuss me personally. I felt supported heard, and valued by my senior leader. It was greatly appreciated.

At the Air Station in Houston, I began to ask how people were doing at home even more than what was my common practice. I asked about their parents and extended family. COVID protocols required people to notify the Command when anyone in their household tested positive for COVID-19 or displayed the common symptoms. I told them I wanted to know if any of their relatives got COVID-19 so I could keep them in my thoughts and prayers as well. I let them know that I understood that the pandemic was affecting everyone in everyone's family and that I cared about them all.

I can't tell you how they personally felt about my actions and questions, but the conversations about how they were doing emotionally and mentally were numerous and routine. They would update me often on when their cousin or aunt got out of the hospital or got out of isolation. They would tell me when they saw me at the air station or they would send me a text. It definitely seemed liked they appreciated the outreach, and I am glad I did it. It was the natural right thing to do, but Intrusive Leadership isn't natural to everyone. Despite the Intrusive Leadership being used in these situations, Intrusive Leadership is not taught at military service academies or leadership training. What Intrusive Leadership is, why it is important, how to do it, and the potential positive impacts aren't explained during courses designed to prepare members for leadership positions. Commanding officers are being directed to use a tool that they haven't been trained on first. That goes against a foundational principle. You can't direct someone without first providing some type of direction. Why is that? I believe the answer is because of current challenges with Intrusive Leadership.

The first challenge is the term itself. The word intrusive immediately turns some people off. It would appear to be a lousy term for a leadership style that I am arguing should be a part of every leader's tool bag. Appearances can be deceiving. In December of 2021, I had the great pleasure of interviewing best-selling author and leadership expert Simon Sinek for my research paper. He agreed that the term was horrible. He actually looked up synonyms for the word intrusive and read them out loud: invasive, meddlesome, prying, and interfering. All of these words have negative connotations and none of them would seem to be appealing for leader.

During the conversation, Simon Sinek did provide a counter argument to his position that the name of the leadership style needs to be changed. *"Nothing describes the actual actions you are advocating that every leader do in caring for their people like the word intrusive"* he admitted. Let me offer this additional argument addressing the "lousy" term.[8] Think about your best friend or close family member. Imagine yourself going through a very difficult situation in your life, a situation that you didn't know how to handle, one that has you on the verge of breaking down. Wouldn't your best friend ask you, *"Something seems to be weighing on you because you haven't been yourself lately, is everything okay?"* Wouldn't your cousin reach out to you and say, *"I know you are going through something right now, and I'm just calling to check on you and see how you are holding up."* Wouldn't you do the same thing if you knew your friend or family member was going through it? We don't look at those situations as meddlesome or prying. We look at them as showing genuine care and honest heartfelt concern. The same actions of Intrusive Leadership, but it doesn't get that same negative response as the term does.

The second current challenge involving Intrusive Leadership is the lack of a uniformly accepted definition of the leadership style. People feel positive or negative about Intrusive Leadership based of what they think it is. What they think Intrusive Leadership is isn't the same across the board however. Major Carl Forsling, US Marine Corps (retired) wrote an article in 2015 titled *"Intrusive Leadership; Bad by Definition"* that was featured in the Marine Gazette publication.[9] In this article he wrote that Intrusive Leadership is "…a bad word that accurately conveys a bad leadership technique that is unfortunately becoming prevalent in the Marine Corps." I had to personally talk to Major Forsling for my research to be complete and unbiased. Fortunately, I was able to do just that in 2022. I sent him my 2020 article to read right before we chatted. Right off the bat, he admitted that what he was calling Intrusive Leadership and what I was calling Intrusive Leadership was not the same thing.

---

8. Simon Sinek, December 15, 2021.
9. Major Carleton Forsling, "Intrusive Leadership; Bad By Definition," Marine Corps Gazette, n.d.

He went on to explain what he thought Intrusive Leadership was and described a situation that motivated him to write his article. The situation he described was being at a unit and being directed to fill out a form prior to weekend liberty. On this form he was asked to provide details of his weekend plans so that the Command could intervene if something concerned them. Was he planning a road trip that included many hours of driving in a short period? Was he planning a social gathering with a large amount of alcohol being served? Major Forsling described the form as a way to provide cover for the Command if something went wrong like an accident or a DUI following the party. The unit members were directed to fill out the report without the Command making an effort to build a rapport. This action didn't go over well with the unit members at all. They didn't trust the Command and the report only made it worse. Imagine yourself going through as he experienced it, but he was a fan of Intrusive Leadership as he define it.

In another article entitled "Intrusive Leadership", Chief Master Sergeant Daniel Tester, US Air Force (Retired), stated that he would have never made it beyond his first enlistment, much less achieved the rank of E-9 if it wasn't for Intrusive Leadership.[11] I had the opportunity to talk to him as well and listen to him describe how an intrusive leader changed the course of his career, his life, and his own leadership style. He described working for a supervisor very early on in his career that would routinely invite him to take a break, have a cup of coffee, and talk. The two of them would actually go and sit in his supervisor's car to get some privacy and talk about life…not work. Chief Tester explained that these talks made him feel valued, it made him feel like his supervisor cared about him beyond his role as an Airman. [12] He felt like his supervisor cared about him as a person. Chief Tester never forgot about this supervisor and how he made him feel. He went on to enjoy a successful military career trying to make others feel that same way. Chief Tester is a huge fan of Intrusive Leadership as he experienced it. Two military members with opposite viewpoints on Intrusive Leadership. Why?

---

10. Major Carleton Forsling, January 31, 2022.
11. Chief Master Sgt. Daniel Tester, "Intrusive Leadership," November 14, 2017, https://www. schriever.spaceforce.mil/News/Commentaries/Display/Article/1371954/intru-sive-leadership/.
12. Chief Master Sgt. Daniel Tester, February 2, 2022.

The International Journal of Environmental Research and Public Health published a 2021 article that discussed a study that concluded that "An intrusive leadership style has been found to be associated with increased absenteeism and reduced productivity, low job satisfaction, and high stress." The study appropriately identified actions that produced negative results for employees. A different journal, Military Medicine, published an article later in 2021 that described Intrusive Leadership as a technique that senior leaders could use as an early prevention technique for at-risk personnel. This article described a leadership root cause analysis that was conducted as a result of a military member at a Naval Hospital having some conduct issues.[13] The analysis identified some missed opportunities for people in supervisory positions to use Intrusive Leadership with this member to identify problems and provide behavioral course correction in a timely manner. Two articles published in the same year, with two opposite positions on Intrusive Leadership. Why?

Due to the lack of a uniformly accepted definition of Intrusive Leadership, people are using the same term to describe different actions. Those different actions yield different results. What Major Forsling experienced is not Intrusive Leadership. What he experienced was simply bad leadership and he associated that negative feeling to the term Intrusive Leadership. What Chief Tester experienced can be considered Intrusive Leadership and he went on to positively impact others with Intrusive Leadership. I believe that there are a lot of people today in the military and civilian workforce who could benefit from Intrusive Leadership, but will never experience it because of the wrong understanding of what Intrusive Leadership is. That needs to change.

The first edition stopped with those two challenges with Intrusive Leadership. Now I would like to add one more challenge with Intrusive Leadership that could be preventing it from being taught as a leadership style. This third challenge is the reluctance of senior leaders to admit that some of the institutional leadership principles that they grew up under are simply outdated and are not going to work for this newer generation. Let me explain.

In the military, there used to be a common saying that was used when service members were facing issues at home with their families. "If the Service wanted you to have a wife (or husband), then the Service would have issued you one!" Ironically, service members were told that problems at home were their business and not to bring their problems to the workplace. Their individual issues were a distraction to the mission and the focus of the unit and being a distraction was never a good thing. This response was the cultural norm back then and everyone knew it. Imagine training and motivating service members to answer the call of duty and be willing to lay down their life for their country. Imagine asking them to sacrifice themselves as a husband, wife, or parent for a cause that is bigger than themselves. Demanding that they give everything they had to unit or mission success, but only want to deal with a small portion of them. Actually leading them as a solider instead of as a whole person. This is simply not going to work today.

I distinctly remember being told very early on in my career that there was no such thing as a personality conflict with your supervisor. It was the responsibility of the junior member to resolve potential personality clashes with their supervisor by changing their actions to appease the senior person. I was told that I should find out what my supervisor likes and begin to like it myself. Learn to play golf if your supervisor golfs. To do whatever I needed to do to make my supervisor like me. Be someone other than my authentic self at the job just so I can achieve career success.

Now, that isn't entirely bad advice. There is some value in understanding what your supervisor has interests in and being intentional about building a rapport with them. However, I do believe there is an unintended consequence to this mindset. Junior personnel who grew up in an organization where they had to change themselves to fit in, turn into senior leaders who think they don't have to do anything to make junior personnel feel included. They end up being the supervisor who thinks trust is a one-way street. Junior personnel must earn the trust of their supervisors and that is it. This is another aspect of leadership that needs to change, and that change is Intrusive Leadership.

In my conversation with Simon Sinek, he also said that I need to find a new term for this leadership style, because it won't be able to overcome the negative feeling that people get when they hear it. That negative feeling will prevent people from adopting the leadership style. I heard Simon Sinek when he said it. I recorded the conversation, so I have actually heard his advice a few times. I am extremely grateful for the time he afforded me to

---

13. David A Paz et al., "Leadership Root Cause Analysis: Using a Formal Analysis Tool to Dissect and Promote Intrusive Leadership in a U.S. Naval Hospital," Military Medicine 186, no. 9–10 (September 1, 2021): 233–35, https://doi.org/10.1093/milmed/usab307.

discuss Intrusive Leadership. He didn't have to, but he did. I hope to thank him for the time he shared and the advice he provided. However, I was not listening.

Like the hundreds of people who were told not to do the milk crate challenge in 2021, I am not listening. I truly believe that with a uniformly accepted definition of Intrusive Leadership, and an understanding of the immense impact that it can have on employees, people will adopt Intrusive Leadership. This is the motivation for this book. This book is me taking the first few steps up those milk crates. When you finish reading, you can decide whether I am standing triumphantly on the other side of the milk crate pyramid, or I took a horrible fall and landed flat on my back.

Let's Go!

PART III

Intrusive Leadership Defined

In order to begin facing the challenges that I described with Intrusive Leadership, it must be defined in clear and understandable terms. Let's do that now.

Intrusive Leadership is a leadership style that uses intentional actions to inspire, motivate, develop, and support people by caring for the whole person through a trusting relationship that addresses professional as well as personal interests, desires, and concerns.

Intrusive Leadership is a leadership style, and it is one of the leadership styles that must be in every leader's tool bag. In my opinion, Intrusive Leadership is the most important style that will be the linchpin to leadership success in the future. It can't be the only style in the tool bag, however. For an organization to succeed, leaders will still have to be technical leaders, servant leaders, transactional leaders, transformational leaders, authentic leaders, and ethical leaders. There will be times and situations that require other leadership styles. The best leaders will know which leadership style to apply to address any situation. Intrusive Leadership doesn't replace the other leadership styles; it enhances them and makes them more effective. Intrusive leaders make better servant leaders. Intrusive leaders make better charismatic leaders.

Intrusive Leadership is an active leadership style. It requires intentional actions aimed at building a connection with people. It takes effort. A leader must make a conscious decision to take certain actions with the sole purpose of building and strengthening relationships with the people that they lead. Asking questions about the individual, learning about their roots, and placing yourself in positions to learn new perspectives, are actions that intrusive leaders take to correctly execute this leadership style. You can't just sit in the office and wait for people to come to you; intrusive leaders take the first step and engage.

Leaders inspire, motivate, develop, and support people. If you aren't doing these things, then what are you doing? You can supervise people without doing these things. Supervisors or managers make sure the work schedule is drafted, tasks are completed, personnel paperwork is filed and approve vacation requests. Leaders do something different. The difference between leading and managing needs to be understood. Managers think about what their people can be doing for them, but leaders think about what they can be doing for their people. Managers want their ideas to be the right way forward, leaders want the right idea to be the way forward. Any manager can have an "open-door" policy, but a leader can create an emotionally safe environment where people want to walk through that open door. There are too many organizations that place people in supervisory or managerial roles and simply expect them to perform as leaders. This can't be assumed. If you want true leaders to be in leadership positions and thrive, then clear expectations should be given, so they execute at the highest levels. Leadership abilities must also be measured. If it isn't measured, then it doesn't matter.

Caring for the whole person is an important piece to Intrusive Leadership. Intrusive leaders understand that they aren't just leading aviation mechanics, combat medics, or marketing analysts. They recognize that they are also leading spouses, parents, and children. What you see at work is a small percentage of who I am despite spending most of my time at work. I am currently a Captain in the US Coast Guard (the best organization on Earth). I am a helicopter pilot. I am also a husband to an amazing Proverbs 31 woman. I am a father to three awesome children. I am a son to tremendously loving and caring parents. I am a brother to a protective sister who still calls me her little brother even though I am in my 40s and almost 6'3". I am an uncle. I am a cousin. I am a friend. I am a Chicago Bears, Atlanta Falcons, Los Angeles Lakers, Miami Heat, New York Yankees, Atlanta Braves, and Georgia Bulldogs fan. Yes, all of them and I have legitimate reasons for each. I am a person that cares about what is going on in my neighborhood, city, state, and country. I have hopes and dreams. I have goals.

Who I am at work might be 10-15% of who I am as a whole person. Intrusive Leadership opens the door for someone to lead more than that 10-15%. An intrusive leader could be a leader for much more. How much more depends on many factors that will be explained later. Intrusive leaders understand that they might only see a small percentage of an employee at work, but that employee is a whole person. Intrusive leaders realize that something could be happening outside of work, that is the underlying reason for performance problems at work.

A trusting relationship is the foundation that Intrusive Leadership is built upon that allows for a leader to lead in a way that has immeasurable impact. This relationship and the importance of trust will be discussed in detail later, but Intrusive Leadership doesn't work unless there is a trusting relationship present and maintained.

Intrusive leaders don't limit their impact to just the workplace. They realize that there are situations that people face outside of the workplace, that affect their performance inside the workplace. They are mindful that their people are human and can't always compartmentalize issues that they are facing on the home front. Intrusive leaders are willing to address professional as well as personal interests, desires, and concerns. They understand how an employee going through a divorce, taking care of an elderly parent, or coping with a special needs kid might need an extension to deliver that job report. They are willing to modify work schedules in certain situations to adjust for special circumstances. Intrusive leaders know which one of their employees is also a personal fitness trainer, loves riding horses, is currently working on the planes but wants to fly them one day. Intrusive leaders know more about their employees and observe for ways to support them in a manner that is best suited for them.

Intrusive Leadership is a leadership style that uses intentional actions to inspire, motivate, develop, and support people by caring for the whole person through a trusting relationship that addresses professional as well as personal interests, desires, and concerns.

This is Intrusive Leadership defined…finally.

"Intrusive Leadership is just good leadership!" is a phrase that I have heard a few times when discussing it with other people in leadership positions. They would go on to say that this is simply what good leaders do, they know their people at an intimate level. I was fortunate enough to discuss Intrusive Leadership with a retired senior military leader who previously served as the Chairman of the Joint Chiefs of Staff. In this role, he was the highest-ranking military officer in the United States Armed Forces. A role that has tremendous responsibility and requires outstanding leadership. When we talked, the former Chairman of the Joint Chiefs of Staff agreed with my thoughts on the leadership style and why it was important to lead in this manner. He went on to say "I never thought about defining it or labeling it because it came natural to me. This is just the type of person that I am and this is what leadership is about." I won't argue that Intrusive Leadership as I define it comes naturally for a lot of people or that this is simply what leadership is about.

However, what comes naturally to some doesn't come naturally to everyone. Also, every person in a leadership position doesn't lead in this manner. If this was simply good leadership then there would be no reason that the US Coast Guard would have to direct it during the lapse of appropriation. There would be no reason for Intrusive Leadership to be directed to Navy Commanding Officers to combat suicide. It would have been happening naturally. Also, isn't servant leadership simply good leadership? Couldn't you say that authentic leadership is just good leadership? The fact of the matter is that there are certain aspects of servant and authentic leadership that need to be intentionally pointed out. That is the same for Intrusive Leadership. There are certain aspects of Intrusive Leadership that need to be intentionally highlighted, further defined, and then discussed. It can't be assumed that every leader is doing it, knows how to do it, or knows when to do it.

I can't tell you how many leadership lectures or courses that I have taken that teach ethical principles when it comes to leadership. The importance of not having inappropriate work relationships, avoiding the temptation of misusing government property or resources, or maintaining a high degree of personal integrity. I could easily say these things come naturally to me. I think they come naturally for a lot of leaders. Yet, we have far too many who have been fired or relieved of their duties for violating the principles of ethical leadership. Senior leaders in the military and corporate America have been caught in inappropriate sexual relationships, using government or corporate resources for personal gains, and more. So, ethical leadership is taught over and over again. Intrusive Leadership needs to be taught as well. There are too many examples in workplaces today where the opportunity to truly connect with employees on a deep and personal level is missed. And when it is missed the impacts could be catastrophic.

Routinely after my presentations or seminars on Intrusive Leadership, I will have people come up to me saying that they never heard leadership explained in this manner. They have encountered some leaders attempting to lead in this manner or either they are trying to, but they never knew what to call it. They struggled to explain it. Now they had a name to a leadership style that they could use and tell others about. That is the powerful impact behind having a definition for Intrusive Leadership. Now that it is defined, it can be taught, explained, and discussed in leadership courses everywhere.

Characteristics of an Intrusive Leader

Trust is the most critical characteristic of Intrusive Leadership and its foundation. A leader cannot begin to develop the type of relationship with an employee that will enable the positive impacts of Intrusive Leadership to be realized without there being some trust between that leader and the employee. An employee who doesn't trust their supervisor will discuss some work issues and concerns routinely but will refrain from discussing personal matters at all costs. While doing my research on Intrusive Leadership, I decided to conduct a quick and short survey with current college students (undergraduate and graduate) and individuals currently in the workforce. In this survey, respondents were asked, "Would you share a personal issue or non-job related concern that might be impacting your job performance with your supervisor/manager?" 16% of the respondents responded "Yes" but 65% responded "Yes, but only if I trusted the supervisor." That is a large percentage of respondents who are indicating they would potentially suffer in silence at the workplace if they worked for a supervisor they didn't trust. Is that okay? As a supervisor, would that bother you? I wouldn't want anyone who worked with me or in my organization to be going through a tough time and not feel like they couldn't trust someone at the job enough to open up about it.

When I interviewed the Director of the Air Force Review Board Agency, Dr. Gerald D. Curry, SES and US Air Force Colonel (retired), he quickly stated "If you really want to be able to reach people when they are down and out, they have to trust you in order to make themselves vulnerable to share whatever it is that is ailing them."[14] I could not agree with Dr. Curry more. A person must trust that if they share with their supervisor that they are having problems with their marriage, struggling with depression, or having financial issues, this privileged information would not be used against them. Their personal situation would not end up being the topic of conversation in the employee break room. They must trust that if they divulge sensitive information, that supervisor would protect them instead of prosecuting them. Admitting that you are having mental health issues can't be viewed as career-ending. This is even the case if they want to share that they have a personal passion that would take them in a different direction then their current professional career track.

---

14. Dr. Gerald Curry, Interview with the Director of the Air Force Review Board Agency, February 9, 2022.

They must trust that they can discuss their desire to work in a different capacity in the organization or even outside the organization, and that desire will be cultivated. Can this employee be given an opportunity to transition to a new job area and re-trained versus being penalized?

Military chaplains enjoy being trusted by soldiers because of their job description. Soldiers know that what they share with the chaplain will stay with the chaplain or be used to seek help for the soldiers. Chaplains typically don't have to earn the trust of the soldiers but other people in supervisory positions do. Just being in a position of authority doesn't automatically translate to being trusted with personal information. This trust must be earned over time and the amount of time will vary from one employee to the next. Intrusive leaders work at building up trust equity with their employees from day one.

For me, I would discuss my desire to support my employees from the very first meeting I had with them during their check-in process. I would ask about their family, their upbringing, and their goals. I would tell them that I would always be available to them if they needed help with a situation at work or at home. I would always say to them, "It is important to me that everything is okay outside of work so that you can give me your best effort while at work. When something is going on that is preventing you from being at your best, I am here to help you through that situation." Now, I understood that just because I said that during the initial check-in meeting, it didn't mean that new employees would automatically start trusting me. It was just the beginning of my effort to build trust so when a moment of crisis arose, my Intrusive Leadership could be effective. Without trust, even the best intrusive leaders won't even be given the opportunity to work their magic.

I want to emphasize that it is the senior person who is required to earn the trust of their people for Intrusive Leadership to work. Not the other way around. I also want to make the point that it doesn't matter what you do as a leader, there are some people who are simply a closed book. They are very private and very reluctant to share personal things about themselves with people at the workplace. That is okay. It is not a requirement of an intrusive leader to get to the point that every person who works for them allows them to lead 30% or 40% of them. Some people are only going to allow you to lead them as an employee and not open up as a father or wife. It is the employee's choice how much of them they are going to allow the leader to lead.

What is required for a person to be an effective leader is that everyone knows that the leader is available and willing to address any concern they might have. Nobody feels like their personal issues do not matter to their supervisor. If at any point they want to share, they know their supervisor will listen intently to learn, and then respond.

It might come as a surprise to you, but I have read more leadership books (audiobooks to be completely honest) after writing one versus before. One of the first books I read after publishing the first edition was "Speed of Trust" by leadership expert Stephen M. R. Covey. This book was recommended when my mentor read this section on trust. I am glad that I took his recommendation. In this amazing NY Times bestselling book, Covey provides a simple and profound definition of leadership in his terms. He defines leadership as "getting results in a way that inspires trust; maximizing both your current contribution and your ability to contribute to the future by establishing the trust that makes it possible." (Stephen M. R. Covey, Speed of Trust, (Simon & Schuster, 2018) p. 40) The book continues to explain why trust is important to effective leadership, what it takes to gain trust, and the impacts of trusting relationships in a wonderful manner. My favorite section is where he describes "Trust Accounts" when building trust with people. Covey lists a few realities concerning these trust accounts:
• Each trust account is unique.
• All deposits and withdrawals are not created equal.
• What constitutes a "deposit" to one person may not to another.
• Withdrawals are typically larger than deposits.
• Sometimes the fastest way to build trust is to stop making withdrawals.
(Speed of Trust, p 135-136)
Trust me when I tell you that this book is filled with nuggets, and I strongly recommend that you add it to your reading list if you haven't read it already. If you are facing a leadership challenge or struggling to connect with someone, ask yourself "do they trust me?"

Emotional Intelligence

Emotional intelligence (EQ) is a key element to several leadership styles and Intrusive Leadership isn't anything different. The concept of EQ has been studied, researched, and written about by multiple people. It is still surprising to me how many supervisors don't have a high level of EQ or people who aren't even familiar with the term. I have conducted presentations on on Intrusive Leadership and have counted raised hands when I ask, "How many people do not know what emotional intelligence is?" Or counted the raised hands when I ask, "How many people have worked for a supervisor with low emotional intelligence?" I have been asked if EQ is something that should be a factor when considering people for supervisory positions.

The term "emotional intelligence" has been around for decades. There are people with doctorate degrees who studied the factors and impacts of emotional intelligence. There are countless number of books or courses specifically on the subject. Harvard Business Publishing listed emotional intelligence as one of the top 10 leadership skills for the future of work.[15] Yet, there's a serious lack of emotional intelligence in many senior leaders. I have seen it, and I have heard numerous stories providing clear evidence of the negative impacts of leaders with low emotional intelligence.

Since there have been many instances of people not being well versed on EQ, I will offer a definition of it, and with additional research into EQ if you happen to be one of these people. Even if I can't convince you to adopt Intrusive Leadership, you will still be a better leader if you adopt a high level of EQ.

Brandon Goleman wrote an excellent book titled Emotional Intelligence where he explains EQ in detail and provides examples of its impact in professional and personal settings. In this book, he states, "Emotional intelligence means being able to pay attention to your emotions and the emotions of those around you while naming or labeling these correctly and to use the emotional information gathered from this exercise to respond appropriately."[16] Now, ask yourself why any hiring manager would not consider EQ when looking at applicants for a supervisory position. Brandon Goleman continues to explain EQ and provides a few competencies of EQ, one of those being empathy.

Empathy is a critical component, and it directly correlates to Intrusive Leadership. In the book, empathy is broken down as well and a definition

---

15. (https://www.harvardbusiness.org/wp-content/uploads/2023/09/CRE4069_CL_RA23_Infographic-3_3.pdf)

16. Brandon Goleman, Emotional Intelligence, 2nd ed. (Brandon Goleman, 2020), 7.

of cognitive empathy is provided. "Cognitive empathy is essentially the practice of seeing things from other people's perspective. Instead of simply processing the information we have from our own perspective, cognitive empathy calls for us to look at that information as if we saw it through someone else's eyes."[17]

Dr. Brene Brown does an excellent job breaking down empathy as well in her book, "Dare to Lead." In this book she states, "In those bad moments, it's not our job to make things better. It's just not. Our job is to connect. It's to take the perspectives of someone else. Empathy is not connecting to an experience, it's connecting to the emotions that underpin an experience."[18] I would also recommend this book to anyone.

How many times have you heard someone pass judgment on someone else for making what they deemed to be a ridiculous decision, having no clue about that person's life story? A person with cognitive empathy understands how a person's life experiences can impact a person's decision calculus. They can understand how a person with the potential to excel in college would choose to immediately start working a low entry-level job after high school, when they learn that person's economic situation requires immediate financial relief. A person with cognitive empathy asks the question "What would cause them to make that choice?" instead of just saying "They made the wrong choice!" Intrusive leaders ask why.

Recently I was blessed with an encounter with Dr. Annie McKee who happens to be one of the leading voices on emotional intelligence. She co-authored a book entitled "Becoming a Resonant Leader" and does a wonderful job discussing emotional intelligence. In this book, leaders who have developed emotional and social intelligence are labeled as effective because "they act in ways that leave the people around them feeling stronger and more capable." (Richard Boyatzis, Frances Johnston, & Annie McKeee, Becoming a Resonant Leader.xx The book goes on to label self-awareness as the foundation of emotional intelligence. Defining self-awareness as "the ability to process emotional information quickly and accurately, to recognize one's own emotions as they happen, and to immediately understand their

---

17. Goleman, 45.
18. (Harvard Business Press, 2008) p. 25)

effects on oneself and on others."[19] This nugget struck me to the core as I looked back on how many times I reflected well after an event. Realizing that my emotions ran wild and ran right over someone else's in the process. Leaving me with an apology to render. That immediate understanding is tough to achieve but a real critical one.

An intrusive leader must have a high level of EQ. They must know when something just doesn't seem right even when a person might say "I am okay." They must know when there is a need to follow up or dig a little deeper. I have been blessed with the opportunity to be mentored by some truly remarkable individuals in my Coast Guard career. Vice Admiral Manson Brown, US Coast Guard (Retired) is one of those individuals. Throughout the years he has given sage council and has provided many junior officers important examples to follow for their career success. When I decided to research Intrusive Leadership, Vice Admiral Brown was definitely towards the top of the list of people that I wanted to interview.

When we spoke about Intrusive Leadership he provided a few excellent points about Intrusive Leadership, how Intrusive Leadership impacted him, and how he incorporated Intrusive Leadership himself. He described this one situation when he was serving as a Commanding Officer of a unit and asked a group of female members, "How are things going at the unit for women?"[20] A specific question with an intrusive intent to learn about a specific aspect of the cultural climate at his unit. Recognizing that something just didn't sound right when they mustered up a mildly positive response, he decided to act.

19. (Becoming a Resonant Leader, p. 26)
20. Dr. Brene Brown, Dare to Lead (New York: Random House, 2018), 140.

He directed the Executive Officer, the second highest ranking person at the unit, to investigate the situation deeper. Recognizing that because of Executive Officer's experience level and leadership abilities, and the need to remain objective, Vice Admiral Brown knew that the Executive Officer was the right person to begin asking intrusive questions. That decision to have the Executive Officer be intrusive yielded a discovery of a senior member at the unit that was sexually harassing the women and was doing so for some time. That member was subsequently held accountable, separated from the service, and unit morale drastically increased. Vice Admiral Brown's emotional intelligence enabled him to take action as an intrusive leader and make the unit a more positive place to be every day for everyone.

Intrusive leaders with high EQ understand the emotions of others, but they also understand their own emotions. They understand that they want to develop close relationships with their employees, but not too close. Some supervisors are working in dangerous career fields and have to ask their employees to put their lives at risk. Military leaders, as well as police and fire chiefs have to make difficult decisions that might have dire consequences without being impacted by those relationships. Despite knowing about a person's growing family or personal desires, getting the mission done has to remain the absolute top priority. Intrusive leaders know how to manage their own emotions to the point that they can care and support the whole person and still maintain the ability to make the correct, yet difficult call.

Committed Effort

Intrusive leaders must be committed to expending extra energy to be intrusive. Building a trusting relationship with someone takes time and effort. An intrusive leader must be willing to stay late to complete required tasks because they have taken time out of the normal workday to engage with their people. There are times that the easy choice is to avoid invasive questions, to take the initial, "Everything is fine," at face value, and not be "meddlesome". An intrusive leader doesn't take the easy way out. They commit themselves to making the extra effort it takes to connect with people on a personal level, to care for them as a whole person, and find out how they can be the best leader for their people. An intrusive leader has lunch with a mentee to discuss career goals, instead of working through lunch so

---

21. VADM Manson Brown, December 29, 2021.

they can leave early at the end of the day. A friend would take the time. A family member would take the time. Leaders need to take the time as well. Intrusive leaders also make the effort to address situations for their employees even when it isn't necessarily required or factored into their own performance evaluation. They do it solely because they care about their people. While I was attending the National War College (Class of 2022, Committee 10, Cherry Parrots, President's Cup Winners), I was fortunate to listen to a Women's in National Security lecture featuring Colonel Cathy Barrington. [22]

At the time, Colonel Barrington was serving as the Commander of the 90th Missile Wing at Warren Air Force Base in Wyoming. She oversees approximately 4,200 people responsible for providing strategic nuclear deterrence by operating and maintaining nuclear ballistic missiles. Missiles that stand on alert, ready to launch at the direction of the President of the United States. Wow! Tremendous responsibility that requires a lot of dedicated work, yet she found herself sitting at the table with local city officials discussing how to make the local community a better place to live for the minority members under her command.

She heard about an African American Air Force member being treated one way when he purchased a firearm at a local store while in uniform, and then being treated in a completely disrespectful manner when he returned in civilian clothes to the same store. She also heard about another African American member having issues with the local school; when his son was called a racial slur and the school's administration only addressed the son's reaction and not the initial treatment of his son. Colonel Barrington decided to act. She decided to approach local officials and began forging a partnership to address the racial climate in the community. These were things that were happening outside the gates of the base she worked.

These situations didn't start when she took her command assignment. Addressing these situations wasn't part of her job description or the reason why the White House would call her office. Yet, she still committed herself to doing something different because it was important her, and important to her members. She knew that she needed all of her employees to give their

---

22. Cathy Barrington, Interview with the Commander of the Air Force 90th Missile Wing, May 5, 2022.

absolute best at work, and that wouldn't be possible if some of her people were being mistreated outside of work because of the color of their skin.

Through Intrusive Leadership, she was able to expose an ugliness in the city that had previously been accepted as standard behavior, something people had to tolerate.

So, what made her commit to the effort it would take to make this impact? Was she directed? Was she taught in a leadership course? Did she witness other senior officers do this? Nope, nope, and nope. Colonel Cathy Barrington made this committed effort to do something to improve the lives of her minority members outside of work as a result of experiences she saw while growing up. She witnessed the treatment of younger siblings, who were multi-ethnic, when the family moved to rural Georgia. The treatment they received and how upsetting it was to her, left a permanent impression on Colonel Barrington. She fought for her younger siblings then, and she is fighting for all of her Airmen now.

And the Airmen are showing her how much her efforts to care for them, in a holistic manner, are appreciated. The feedback given for her making the effort to improve things in the community has been remarkable. The impact has been far reaching as well. Other units in other states have reached out to Colonel Barrington for guidance after similar racial situations were discovered in their surrounding communities. Now Airmen throughout the Air Force are being cared for better, all because of the committed effort of an intrusive leader. An intrusive leader that has now been selected for General.

Active Listener

Intrusive leaders must have the ability to listen. They must be able to listen to their people, hear what is truly being said, and at times hear what is not being said. Active listening is the practice of preparing to listen, observing what verbal and non-verbal messages are being sent, and then providing appropriate feedback for the sake of showing attentiveness to the message being presented. Active listeners engage in conversations in a manner that conveys to the other person that they are being heard. Do not underestimate the impact on someone when you are able to provide them that sense of being heard when they are going through a difficult moment.

One of my most challenging and rewarding assignments in my Coast Guard career was serving as the Military Aide to the Commandant of the Coast Guard. In this role, I had a front row seat watching Admiral Paul Zukunft lead the organization for two years. Every time Admiral Zukunft had to engage with Congressional members, meet with shipbuilders, exercise diplomacy with foreign heads of state, and address the members of the US Coast Guard, I was right there. My specific role as part of the team that directly supported him was more logistical than informational. I was the person that made sure the Admiral had received his daily binder or classified folders, not the person responsible for what was in the binder or folder. I made sure we were in the right uniform, headed to the right location, at the right time, for him to carry out his vitally important duties.

Looking back at the two years I was fortunate enough to travel the world and have some amazing experiences serving in this position, one of the things I remember most about Admiral Zukunft was his ability to make me feel heard. Here was a man who routinely discussed important matters with the top leaders of the United States. Yet, he still listened intently as I discussed my feelings as an African American watching the riots in Baltimore, Maryland after the death of Freddie Gray.

We talked about my experiences growing up, my time at the US Coast Guard Academy, being a newlywed, and becoming a father. Admiral Zukunft, while preparing for Congressional testimony and engaging with Coast Guard members, took the time to listen and learn about me. There were times that I would listen to him speak and he would reference something that we discussed months earlier. I remember telling him in passing about a book that I had recently read. That weekend, he bought the book, read it, and came into work that following Monday looking to discuss it (Sidenote: it took me months to read the same book).

Routinely, while driving to or from important meetings, Admiral Zukunft asked about my thoughts on a topic, and he listened. Often, that topic had to do with life, not just work. It didn't matter if he agreed with me all the time, or always took my recommendation about a policy matter when he asked for it, he made me feel heard each and every time. That sense of being heard was powerful. Admiral Zukunft was an intrusive leader, and his superpower was active listening.

Leadership expert John C. Maxwell has written a number of New York Times best-selling books on leadership and has been given the label as America's #1 leadership authority. In his book "Good Leaders Ask Great Questions", he discusses the power of listening. He explains how listening demonstrates that a person values others, and he provides a memorable quote from another author, David W. Augsburger. "Being heard is so close to being loved, that for the average person, they are indistinguishable." [23] When I think about how Admiral Zukunft made me feel by simply listening. Or when I think about my other favorite leaders asking for my opinion on important matters even though my job title at the time did not offer me a seat at the table where decisions were being made. And when I think about leaders who didn't. I would have to agree with David Augsburger's statement.

While doing a presentation in California at a military unit, a senior enlisted member stood up from his seat and provided a story to support my message to those in the audience. He explained that he had already heard about my message on Intrusive Leadership through another co-workers and he decided to make a change in how he led. He decided to be very intentional in making sure people knew that he was listening to them. When he asked other military members how they were doing, he made sure he faced them and listened intently. He went on to say that one day he was approached by a young junior enlisted member at the unit because he wanted to talk. This junior member started the interaction by saying "Senior Chief, you are the only person that when you ask me how I am doing? You actually listen and I am not doing okay right now, and I need some help." Senior Chief created a space to positively impact this individual's life.

There are plenty of books out there that can provide you tools and instructions on how to be an active listener. I really enjoyed Heather R. Younger's book entitled "The Art of Active Listening." She is a best-selling author and a leadership expert. In this book she breaks down active listening and walks you through her 5-step model that provides a playbook. I highly recommend this book to you if you feel that you can improve your active listening skillset.

---

23. (John C. Maxwell, Good Leaders Ask Great Questions (Center Street, 2022) p.51)

Transparency

Intrusive leaders must be transparent. They must let people have a peek into their world, thoughts, and challenges if they expect people to reciprocate. One of the ways leaders can break down barriers and build some trust is by first opening up and talking about themselves. The flow of information must go back and forth. If a person is going through marital problems, open up about your challenges if you had similar experiences. If a person is weighing career options, discuss how you worked through situations and how you made career choices. Being authentically transparent can open the door to developing a trusting relationship with an individual who would typically choose to remain private. It came as no surprise that Stephen M.R. Covey listed transparency as the third behavior of the 13 behaviors that leaders should display to gain trust. Saying that transparency will not only build trust, but it will build trust fast. [24]

Often, leaders are not looked at as regular people with normal personal challenges. We often place leaders on a pedestal. The pedestal where all the perfect people who always excel in everything and have never faced any adversity reside. When you place someone on this pedestal, it is tough to consider them as someone who would also understand what you are going through. Intrusive leaders must take themselves down from that unearned pedestal by being transparent in order to be seen as someone who would understand the challenges that other people are encountering.

During this past year, I got to know Chief Jason Kuzik who has recently retired after faithfully and honorably serving in the U.S. Army and the U.S. Coast Guard. He is an intrusive leader, and you are going to hear an amazing story about him later but he did provide me with a great example of transparency from a leadership perspective that he allowed me to share.

One day while working as the Officer in Charge for a Coast Guard small boat station in the Great Lakes, Chief Kuzik was sitting down and speaking with one of his members. This conversation centered on life and family, something that was routine for Chief Kuzik to do with everyone. During the

---

24. (Speed of Trust, p. 157)

conversation, the member decided to open up and discuss an issue he was having. The member, a young male, discussed that he and his wife were trying to have a baby but were encountering problems conceiving. He suspected that the issue was with him and was concerned. That is something very personal to share with family and friends, not to mention a supervisor. How did Chief Kuzik respond? He opened up and was transparent about his issue with low testosterone levels. He told the member how he was in a similar situation, went to the clinic, and got some help. He recommended that the military member do the same. He didn't judge, he didn't laugh, and he didn't decide to conceal his own experience. His transparency strengthened an already present trusting relationship. The member listened and followed Chief Kuzik advice to go see a doctor. A few months later the member returned to that same office to tell his supervisor that he and his wife were expecting their first child and couldn't be happier.

Transparency has impacts at the organizational level as well. I believe that nothing strengthens organizational trust like transparency, even in the midst of a crisis or scandal. An organization can work through a damaging situation and come out even stronger if the organization decides to be transparent throughout the process. People can accept mistakes, and provide some grace by offering opportunities for recovery, but that can't happen without transparency first. It is tough for people to move forward if there is an ongoing effort to avoid being transparent about bad decisions being made or bad actions being taken. Leaders can't demand transparency up the chain if they aren't willing to display transparency down the chain. That lack of transparency will remain a roadblock to organizational unity and loyalty. These five characteristics were the only characteristics that I mentioned in the first edition. Since then I have seen the need to add two more. These characteristics have constantly come up again and again this past year needs to be called out. So, I am adding them now.

Humility

Humility closely relates to transparency, but I do think there is a big enough difference that it needs to be separated. The Webster's definition of humility is freedom from pride or arrogance; the quality or state of being humble. I'm sure you can agree that nobody likes to work for someone who makes everything about them. Leaders that care more about being right versus getting it right. Leaders who are unwilling to admit mistakes or overstate their importance. Leaders who won't reverse course on a decision they made after there are clear indicators that decisions are producing negative results. Leaders that overstate their importance or falsely assume that everyone working for them aspires to be them.

These leaders are tough to work with or for. They take the air out of the room. They diminish creativity, limit discussions, rarely seek recommendations, and struggle to inspire anyone. Unfortunately, their lack of humility probably blinds them to the reality of their negative impact. Their position or title means they have achieved success so there is no need to question their abilities. They attribute their success solely on their hard work and sacrifices so they see the need to offer mentorship to junior personnel. Leaders who aren't humble oftentimes negatively impact morale and employee satisfaction.

Every great leader that I have had the pleasure to encounter has led with humility. I always knew they were different but that connection between humility and great leadership became crystal clear when I read Jim Collins' best-selling book "Good to Great." Remember, I already mentioned that I read more leadership books after writing the first edition than before. Hard to believe it took me so long to read this gem, but I am here now. Stop judging me!

In this book, Jim Collins and his team analyze leaders and successful companies to discover what is the difference between good companies and great companies. The difference hinges on leadership, and they discovered that all companies that made the transition from good to great were led by a leader with one common trait. That trait was humility. These leaders who were labeled Level 5 leaders were in great companies and displayed an extreme level of personal humility. "Level 5 leaders channel their ego needs away from themselves and into the larger goal of building a great company. It is not that Level 5 leaders have no ego or self-interest. Indeed, they are incredibly ambitious – but their ambition is first and foremost for the institution, not themselves." [25]

The humility within intrusive leaders, mixed with ambition, drives their will to succeed to be able to do more for the people they serve. They want to have more impact as a mentor. They want to have more impact as a mentor and be considered a sponsor. They want to earn a seat at the table to be a voice

---

25. (Jim Collins, Good to Great, (HarperCollins Publishing, 2001) p. 21)

for those that aren't in the room. They work to be the person at the head of the table to be able to provide seats at that table for others. They simply are driven by the thought of being able to do more for more people. Who wouldn't follow that type of leader?

Humility also enables intrusive leaders to ask the important question during a counseling session, "How can I be a better leader for you?" They always have the opinion that there is something that they can learn about leadership and those lessons can come from anywhere. They are continuously trying to improve and become a better leader for the people they are trying to lead. The moment you think you have nothing else to learn about leadership and leading people, a humbling experience is on the horizon.

I will be the first to admit that I have made leadership mistakes, and I will probably make more in the future. I have learned a lot about leadership from the very people I was assigned to lead. I do consider myself an expert on Intrusive Leadership while also understanding that there is more to learn. There is more research to be done and questions that deserve deeper probing answers. This is a humble journey that I plan to continue, and I am honored that you are here with me.

Courage

The last characteristic of intrusive leaders is courage. That might seem like common sense or an obvious trait, but I do think that courage should be explained from a different perspective. I am not referencing courage from the perspective of running into a burning building or confronting fierce enemies. It takes courage to pursue a high-profile assignment where the margin for error is non-existent, and failure is career-ending.

The courage that I am talking about is the type of courage it takes to have a difficult and sensitive conversation with an employee. To ask specific questions that might lead to an uncomfortable conversation. Questions like *"How is your experience here working with us as a woman?"* or *"How are you coping with being a law enforcement officer and being African American in this moment?"* or even "You don't seem to be your normal self these past few days, are you okay?"

Many leaders and some employees may not have the courage to ask probing questions because they are afraid of the penetrating response. They are afraid that they will find out something that they will have to act on as a leader.

They are afraid that they might offend the person. They are afraid that they are opening themselves up to a conversation about something that they don't necessarily have the answer to. More than a few people mentioned to me that they intentionally avoided asking about how their co-workers were doing during the racial tension in 2020 because they simply didn't know what to say to fix it. They didn't have an answer to racism or a confident response that would reduce mental confusion. So, they chose the easier more comfortable path. They chose to say nothing and let their co-worker suffer in silence.

Intrusive leaders don't run from these moments, they recognize them and decide to be courageous and ask the question. They are willing to be uncomfortable in order to put themselves in position to support someone else. I will tell you that when I was going through my moments of being mentally "not okay," I didn't need someone to have the answer. I just needed someone to recognize my pain and simply listen. Intrusive leaders would rather be uncomfortable and engage in a sensitive conversation versus letting one of their people suffer in silence.

Some leaders don't want to ask questions about workplace harassment or discrimination because they know they will be forced to act. They would rather float around on a cloud of illusion that everything is perfectly fine within their organization, than having the courage to face reality. It might sound crazy, but there are people in leadership positions who would rather stay oblivious to certain toxic activity than being forced to hold someone accountable. I have heard the stories and unfortunately, I have seen it with my own eyes. One of General Colin Powell's leadership lessons hit this right on the nail. In his list of 18 leadership lessons, the very first one is *"Being responsible sometimes means pissing people off."* [26]

---

26. (Curt Stowers, 18 Leadership Lessons From Colin Powell, 12/11/14, https://www.f5fp.com/18-leadership-lessons-from-colin-powell/)

Intrusive leaders want to know about toxic behaviors because they want to put an end to them. They might have the courage to hold someone accountable even if it means holding a senior person accountable, maybe even a peer. They ask if someone is okay because they would rather ensure the collective efforts in getting that person some help versus an unfortunate alternative.

Intrusive leaders have the courage to ask difficult questions. They have the courage to enter uncomfortable spaces. They have the courage to engage on sensitive topics. And they have the courage to take action when it is deemed necessary. Intrusive leaders don't let fear get in the way of making their people feel valued, supported, and protected.

These seven characteristics aren't the only characteristics of a good leader who leads in an intrusive manner. The characteristics of Intrusive Leadership overlap and intersect the characteristics of servant, authentic, ethical, and charismatic leaders. Integrity, being ethical, and selfless, are additional characteristics that intrusive leaders have. However, these characteristics are important enough and have a specific impact for Intrusive Leadership. So, much so, that they needed to be pointed out and intentionally discussed.

PART IV

Is Intrusive Leadership What The People Want?

Now that Intrusive Leadership has been defined and the critical characteristics of an intrusive leader have been discussed, I want to address an important question. Is Intrusive Leadership a desirable leadership style? Is it what the people want? No… Intrusive Leadership is not the leadership style people want. It is the leadership style that people NEED! People don't just want to feel valued and supported, they need it. They need to know that someone cares for their well-being, cares about their mental health, and cares about their dreams. Some people have said to me that the younger generations don't want people intruding into their personal lives. They say this when describing the same generation that tells the world what they had for dinner, where they went on vacation, and what new dance move they learned all over social media.

During my presentations on Intrusive Leadership, there are people that initially have negative thoughts on the term and don't want to be led by an intrusive leader. By the end of the presentation and having an understanding of what Intrusive Leadership actually is, they indicate that they have changed their mind and want to be led by an intrusive leader as well. There are mountains of evidence that proves far and beyond any reasonable doubt that Intrusive Leadership is desirable.

While I was in one of my seminar classes at the National War College (NWC Class of 2022, Committee 10, Cherry Parrots, President's Cup Winners), the instructor opened the daily session by asking each student to describe their favorite leader that they have worked for. Almost every student described an intrusive leader. They told stories about leaders that went the extra mile for them while they were pregnant, leaders that made sure they got extra time off after just getting married, and leaders that helped them maintain a healthy work-life balance. They didn't describe leaders that were the best pilot they flew with, the best negotiators working on an international trade deal, or the best military tank driver. They described leaders that made them feel the best as a person. Some of my classmates had never heard of Intrusive Leadership, but they had already benefited from it.

In January of 2022, I was very fortunate to be able to talk about Intrusive Leadership with General CQ Brown Jr. At the time he was serving as the Chief of Staff of the US Air Force, the highest-ranking military member in the Air Force and a member of the Joint Chiefs of Staff. He told me that when he travels the world and meets with Airmen, he routinely hosts a *"Breakfast with Airmen."*[27] At this breakfast, he sits with 10-15 junior enlisted Airmen and discusses what is on their minds. He mostly listens to them and answers their questions of him. The only thing that is scripted is his question to the Airmen about leadership. The General simply asks them, *"What is the main thing you want from your leadership?"* The response that is most often given is that the Airmen want their leadership to get to know them as a person and they want their leadership to care. What they are saying without using the term, is that they want Intrusive Leadership.

When I think about the leaders that have had the most significant impact on me, the leaders that helped me through difficult situations, the ones that have provided the necessary feedback to improve my performance, they all were intrusive leaders. They all knew me personally and knew about my family. We have engaged in numerous discussions about life, being a parent, and more.

I remember one particular situation in 2015, when Captain Aaron Roth, US Coast Guard (retired), asked if he could take me out for dinner. At the time, he was in charge of the Commandant's Advisory Group while I was serving as the Military Aide to the Commandant. Even though, Captain Roth was not part of my direct chain of command, we were both part of a larger team that worked to support the Commandant of the Coast Guard. Since he asked to take me to dinner, I knew he was paying so it took no time at all for me to respond with *"Yes sir! When would you like to go?"* We set the date and time and I showed up prepared to discuss my performance, my desired next assignments, and my career aspirations. I saw this dinner meeting as an opportunity to gain another mentor at work that could help me further my career. I was ready.

---

27. General Charles Q. Brown Jr., Interview with Chief of Staff of the Air Force, January 21, 2022.

When we sat down, we ordered drinks and our entrees. After the waiter took our orders, Captain Roth kicked off the professional development session by saying "Marcus, in a few months you are about to have your first child and I want to talk to you about how that is going to change your life." I wasn't ready. This was a personal development session by a senior officer at the workplace to make sure I was prepared for a life changing event outside of the workplace. I didn't just gain a work mentor; I gained a life mentor.

Captain Roth retired from the Coast Guard after his time supporting the Commandant, but we still remain in contact today. We talk about what is happening with my family, my career, things that are happening in society, and our beloved Chicago Bears.

I confidently stand firm on my declaration that Intrusive Leadership is not only desired by people in the workplace, but also what they need. Employees need to feel seen, heard, and cared for. The response that I get from young people in college or beginning their professional careers is that they want to work for an intrusive leader. Even when they are confused or concerned when hearing the Intrusive Leadership term for the first time, when it is explained to them, they express the desire to experience it. It is time to listen to them and give them what they want.

While I was writing this book, I was completely surprised with an opportunity to talk to Captain Sully Sullenberger. Yes, that Sully. The pilot who famously landed a US Airways airline jet on the Hudson River in 2009 after a bird strike disabled both engines shortly after takeoff. The perfectly executed emergency water landing saved the lives of all 155 passengers and crew aboard. Simply remarkable and I got the chance to talk to him about Intrusive Leadership. We talked about several things but what impacted me the most was our discussion about when he needed Intrusive Leadership. While he was working for an airline company, he tragically lost his father to suicide. I could feel the pain that is still with him today as he described that moment in his life. Sully went on to discuss losing his mother three years later to cancer. A few months after being diagnosed, she was gone. In both situations involving losing one of his parents, he was given four days off from work. Imagine that…four days. "That was the policy back then," he said.[28]

---

28. Captain Sully Sullenburger, September 8, 2022.

I asked him what impact it would have had on him if someone at the job would have reached out to him and said something like, "Take all the time you need. You are going through a difficult time, and it is more important for you to be the son your mother needs right now as your family deals with the loss of your father. Be there for her. We will be fine until you return." Sully said it would have made a world of difference to him. Unfortunately, nobody made that call and said those words. Four days to process the loss of a parent and then back to work. Sully came back to work both times and did his job. That was the culture of the workplace back then.

Times are different now. This younger generation isn't coming back if a job shows little care for them when they are in a time of need. Even if you do see them after four days, you will be watching them quietly quit until they find a place that suits their personal needs better. They won't be the person you need when the company needs a miracle to be performed. They will always remember when the company wasn't there for them. The way Sully was treated was never okay and it should have never been the policy. This generation is different. They aren't going to simply accept that policy and show back up. They are different and how we lead them must also be different.

As I mentioned earlier, after a year of the first edition being available and many more presentations completed, I am even more convinced that Intrusive Leadership is highly desired. Time and time again I have had people request that the presentation be done for the senior managers at their company or mentioned how the leadership style would have positively impacted the career of a coworker or themselves. I have heard stories of someone needing a leader that truly cared about them, and that opportunity being missed. One year later and the need for Intrusive Leadership is becoming more and more apparent.

Intrusive Leadership In Action

What does Intrusive Leadership look like? What are the actions of an intrusive leader?

What I am going to do now is simply provide some examples of Intrusive Leadership in practice and in action. I am going to describe a few situations of different ways that intrusive leaders have used Intrusive Leadership in order to foster trusting relationships with their employees. What you won't read is a well-defined "How-To" guide on how you do this. Sorry it isn't that easy. I'm not going to give a step-by-step playbook for you to memorize and then execute. Intrusive Leadership is a personal leadership style that is best done by drawing on the personal strengths of the leader and the needs of the employee. The method in which I do Intrusive Leadership works for me, but it might not work for you and vice versa. How I connect with one employee might not work for another employee. The "how" for Intrusive Leadership might change from leader to leader and situation to situation. The "why" for Intrusive Leadership will not.

Example 1: Intrusive Leadership from Day One

While I was the Commanding Officer of Air Staion Houston, I would meet with every new member as part of their check in process. During this meeting I would set the expectation of what I needed from the new member being a part of our team. I let them know that I expected their absolute best and their maximum effort to execute the mission safely and effectively. I also set the expectation of what they will be getting from me. They would get my absolute best and my maximum effort in supporting them. I always had my Command philosophy printed out and I would hand it to them. I would point out the first section entitled People. In that section was written:

People
We will be excellent in our care for each other and our families.

The Coast Guard's most important resource is its people. Every single member of this unit represents a specific capability in the service's inventory. The aircraft, facilities, and supporting systems do not accomplish the mission. The mission is accomplished by the people and that cannot happen unless we care for our people first. We will make sure that support is provided, training opportunities are awarded, and family concerns are addressed. Nobody should ever question how much he or she is valued!

When I discussed this section with a new member, I let them know that I would do my absolute best to care for them as well as their families. I told them that it was important to me that their home life was good so that when they were at work, the task at hand had their undivided attention. If they were ever in a situation where things were not good and that their performance at work was being impacted, then we needed to know so we can help. I told them that this was the same direction that I gave their supervisor, so they had a direct line to me if their needs were not being met.

Then we would talk about them as a person. Where they were from, if they were married, kids, where have they been stationed before, what were their career goals, hobbies, interests, personal goals, and more. I tried to get as much information about them as I could so I could use it later to create a personal connection with them. From the very first encounter with a new member, I let them know that they should expect to be led in an intrusive manner.

I wrote my Command philosophy in 2019, well before I would become a champion for Intrusive Leadership, but many years after being impacted by it. Can you hear the Intrusive Leadership being spoken even without the actual words being spoken?

Lieutenant Colonel John Wallace had the same approach as I did when he was serving as a Commanding Officer in the US Air Force. He started developing trusting relationships with his members during their check-in meetings with him. He explained to me that this technique paid huge dividends as his unit responded to personnel concerns following the murders of George Floyd, Breonna Taylor, and Ahmaud Arbery.

Captain Adrienne Waters had just recently reported to the unit when she had the strong desire to do something in this moment of extreme racial tension in our country. As a result of the trust that was built during that first meeting with her new supervisor, she felt comfortable approaching Lieutenant Colonel Wallace with an idea to take action.[29] The idea that she had was to hold multiple training sessions at the unit discussing topics like racism, unconscious and systematic biases, and the experiences of African American members in the Air Force.

These training sessions discussing difficult and sensitive topics were attended by all members of the unit and played a major role in broadening the perspectives of everyone involved. What enabled Captain Waters to have

---

29 Lt Col John Wallace and Cpt Adrienne Waters, December 15, 2021.

the courage to approach her boss with this idea that some would consider as risky? "I trusted him. Even though I had only been working for him for about a month, he had built up enough trust equity with me that I felt like I could come to him, and he would at least listen. Our very first conversation, he carved out time to talk to me about me. We didn't just talk about operations or the job. We talked about who I was. That trust built from the first conversation by an intrusive leader, enabled her to be the courageous leader that their unit needed at a critical moment.

Example 2: Intrusive Leadership During Counseling Sessions

Admiral Michelle Howard made it a habit of asking a very specific question when she conducted performance counseling sessions as a supervisor or Commanding Officer in the US Navy. She would ask her members, *"Is there anything going on with you that is preventing you from being your absolute best at work?"*[30] This question would illicit answers that dealt with various things at work.

One member expressed how he struggled to manage his email account and it caused him to spend precious time searching for work tasks and other things. This led to a meeting where Admiral Howard helped him organize his inbox and created different folders to enable him to group emails together according to relevant topics. This led to increased work efficiency. This member came back to Admiral Howard a few weeks later and explained how organizing his email folders freed up hours of work time every week.

Other answers to Admiral Howard's intrusive question dealt with issues outside the workplace. Another subordinate of hers explained how the current living situation of a close family member was causing a lot of stress for her. The living situation was one that caused this subordinate to constantly be concerned about the family member's well-being and exposure to certain things. This answer caused a discussion about what options this subordinate had to positively impact the situation and what would relieve the stress it was causing. This discussion produced a solution. Admiral Howard recognized that the cause of the stress was the subordinate's inability to communicate directly with the family member on a routine basis and that caused days to go by where she didn't know if the family member was okay. Admiral Howard suggested that the subordinate buy the family member their own

---

30. Admiral Michelle Howard, December 29, 2021.

cellphone under a family plan to enable them to communicate directly at any time. The subordinate took the suggestion and purchased the phone. A few months later Admiral Howard checked in with the subordinate and found out that there was no more stress, and the relationship between subordinate and the family member was stronger than ever. A solution that addressed a concern outside the workplace that was brought to light from an intrusive question during a counseling session within the workplace.

I have recently made a change in how I conduct counseling sessions since I began researching and advocating for Intrusive Leadership. I specifically ask an intrusive question with the intent of driving home the point that I care about the individual as a whole and not just the employee. I knew it was important to ask the question, but I was shocked by the responses.

During counseling sessions, I would have a junior member explain how everything was going right for them at the job. They were receiving good performance marks, they enjoyed the assignment they were currently working and had future goals that they were working on achieving. Everything seemed to be going great and they were content leaving the conversation right there.

After a while, I would shift the conversation to a simple question. *"We have spent the last 15-20 minutes talking about LT Smith (the employee) and it seems like everything is going great for LT Smith. How is Chris (the person) doing? How are you doing outside of work?"* A few times after I asked that question, the person who was just explaining how their work life was going had a change in emotions. Things were going great at work, but they were not going great outside of work. Something was going on in their personal life that was causing them some pain or mental anguish. Asking them that question in that manner, catching them off-guard, revealed something that they were currently struggling with in silence.

A few times we began to discuss what personal issues they were having. They needed to talk to someone about it. In either case, I would end the session by saying this. *"I want you to know that I like LT Smith but I love Chris, and I need Chris to be good so LT Smith can be great. If you ever want to come and talk about what is impacting Chris, then I will always make myself available to you. I never want you to think that I only care about you as an employee. I care about you as a person."*

Example 3: Intentionally Being in the Right Place to Connect

My first aviation assignment after earning my *"Wings of Gold"* in October of 2003 was Air Station Miami, Florida. It was an awesome assignment where I was challenged routinely and learned a lot about myself and operational

leadership. Conducting aviation missions in the Straits of Florida and deploying on Coast Guard cutters (The Coast Guard calls its ships cutters) was a great experience. I had my first life saved, first shipboard landing, first rescue off a cruise ship, and deployed to New Orleans to take part in the Coast Guard's response following Hurricane Katrina. So many memories made and so many relationships built. One relationship stands out. That is the mentoring relationship that developed with Captain Todd Lutes.

During this time in Miami, I played basketball a lot. I played in local city leagues, I played pickup games at a nearby LA Fitness gym, and I played at lunchtime at the Air Station. During lunch, a group of us would routinely meet up at the basketball court, even underneath the sweltering summer sun, and play a few games. Most of the guys playing were the young pilots and some of the junior enlisted members. Usually everyone was under 30 years old, and many of us playing were minorities. There wasn't that many of us minorities at the unit, but this was our opportunity to connect with each other playing a game that we all loved since growing up in different neighborhoods throughout the country.

The normal list of players got thrown for a loop one day when the Operations Officer showed up to play. To our surprise, then Commander Todd Lutes joined us to play basketball. I remember looking at his shorts, that didn't come down to the knees like the cool kids and began talking to one of the other players. One of the other guys asked me, "Are we supposed to let him win?" My response was quick. *"Hell no! He has to earn everything he gets."* Real ballers know the rules…take it easy on nobody.

So, we played a few games and the lunchtime basketball session ended with Commander Lutes going winless. He played hard and held his own but what surprised us even more was that about a week later he came back out and played again. What was he doing? He wasn't winning games and you could see him walking around the unit a day or two after playing in some slight pain. I didn't immediately recognize it then, but what he was doing was intentionally putting himself where we were to build a connection. Amongst all the rebounds and pick and rolls, we cracked jokes and gave out high fives. We did what we normally did but we did it with one of the senior officers at the unit. This was different for us.

We weren't used to a senior officer coming to where a group of minorities were to connect with us. We were used to having to go to them. We had to attend the holiday parties, play on the unit softball team, and participate in the morale golf outings. We had to make sure we put ourselves in position to connect with senior members and take the initiative to develop relationships like the other pilots had.

Through losing basketball games, Commander Lutes won us over. Some of the benefits of the relationship that Commander Lutes built will be discussed later but I will tell you that the relationship continues today. Years after I left Miami and I found myself nervously preparing for an important special assignment interview, the first person I thought to call to calm my nerves was Captain Lutes. He answered my phone call, gave me some encouragement, calmed my nerves, and I got the job. I routinely sent him my performance evaluations for honest feedback and discussed future job assignments.

When he encountered a situation involving a member that had a license plate frame that included a Confederate Flag, he called me to discuss my thoughts on the situation and the actions he took to address it. Throughout the years we have had numerous conversations about my career and my life. We have had deep discussions about sensitive topics like racism without any fear of judgment. When I think about senior officers that I have been fortunate enough to call true mentors, my relationship with Captain Todd Lutes is high on that list.

Intentionally placing yourself into positions to connect with different people and learn about their experiences and perspectives to craft a leadership style that is effective for everyone, is something that I learned from Captain Lutes. When I was assigned as the Operations Officer at Sector/Air Station Corpus Christi, Texas, I put what I learned into action. I made sure that I attended the Women in Aviation International conference in Orlando, Florida. This is an amazing conference that is dedicated to bringing women from all over the world together to honor women trailblazers in aviation, discuss current issues and opportunities in the aviation industry, and motivate the younger generation to seek an aviation career. I got the opportunity to hear stories from fellow Coast Guard pilots about their struggles to be included as a part of the team at the same level as their male counterparts.

I recall one woman who shared a story about when she was serving at an operational unit and all the pilots got together for "Guy's Night Out." This was simply an evening where the pilots played poker, drank alcohol, and cracked jokes.

A good time connecting with one another, building team unity, that even included the Commanding Officer. It included everyone except the women pilots. Just imagine how that could make someone feel. I have been alone at a unit as the only African American pilot, but I was still invited to hang out with the guys. This woman however, surrounded by teammates that she had to conduct dangerous lifesaving missions with, was placed on an island during "Guy's Night Out." I heard other stories.

I heard about the challenges of serving and being a mother. I met a few junior women pilots just starting their careers and forged mentoring relationships. I also witnessed a lot of men there talking to airline company recruiters trying to put themselves in position to obtain a commercial airline job. They were taking advantage of the employment opportunities at this conference for women, while completely skipping out on the opportunity to learn from them. It was a tremendously rewarding experience for me that I still benefit from years later.

Captain Lutes also provided me with one of those "scared straight" moments. Scared Straight is a community service program where the focus is to steer troubled teens towards a positive path by scaring them during a visit to an adult prison. My moment with Captain Lutes doesn't include an adult prison but there was some genuine fear. One day, while assigned to Air Station Miami, I volunteered to support Career Day at a local elementary school.

Getting out in the community, connecting with the public, and sharing information about the US Coast Guard. Here I was taking time to give back to the public I love and Captain Lutes saw a moment to seize upon. As I got out of my car and started to walk toward the main entrance to the school, two kids darted around some cars running towards me screaming "Daddy, Daddy, Daddy!" All breathing and heart pounding stopped. I was frozen in fear. My knees started to weaken. As my innocent world started to crumble and consciousness started to fade away, I noticed that my surprise offspring didn't have the type of hair or skin tone that I would have expected.

There was no melanin in these kids. Something wasn't right. What wasn't right was Captain Lutes, sitting in his car parked a few spots away, laughing uncontrollably. Using his kids to scare the living daylights out of an innocent soul. This is not an example of good Intrusive Leadership, but this story must be told. #NeverForget

Example 4: Light in the Midst of Darkness

It is easy to connect with someone on a good day or high moment in their career. It seems natural to congratulate someone on a promotion or when they received some formal recognition. However, Intrusive Leadership is vitally important in those low moments or tough days. When someone has made a mistake, exercised poor judgment, or has been accused of wrongdoing, the tendency is for leaders to separate themselves from that individual. There is a natural desire to not engage for the fear that associating with that individual would in some way have negative impacts for the senior official reaching out. At a time when support is most needed, alienation often occurs.

Admiral Thad Allen, the 23rd Commandant of the Coast Guard, was not the type of leader that separated himself when someone made a mistake. Widely known for his leadership in the aftermath of Hurricane Katrina where he served as the lead Federal On Scene official. Conducting daily nationwide interviews and coordinating the largest integrated lifesaving mission in US Coast Guard history. He stepped up and displayed remarkable leadership in response to a historic storm, and he did the same thing when people went through personal storms.

When I interviewed Admiral Allen about Intrusive Leadership, he described it as an active leadership style in contrast to passive ones.[31] He explained that passive leadership styles wait until someone sends out signals to get information. Looking and listening for signals to gain understanding. Recognizing that passive leadership doesn't always provide the full picture, intrusive leaders actively send out signals to get a return signal to get information that they wouldn't get any other way. They ask questions to solicit a response that will give them valuable information to allow them to make a connection and eventually make a difference. He went on to describe specifically being there for people when most people wouldn't.

---

31. Admiral Thad Allen, Interview with the 23rd Commandant of the Coast Guard, January 26, 2022.

One situation he described was when a Commanding Officer of a Coast Guard cutter was being investigated as a result of a pretty significant at-sea collision with another vessel. This officer previously worked on Admiral Allen's staff so he reached out when he didn't have to, but knowing it was needed. The Commanding Officer was facing the potential of his career coming to an end as a result of the mishap, but Admiral Allen gave him some encouragement. I was not fortunate enough to interview that officer on the receiving end of that encouragement, but I can only imagine the impact that it had.

In the US Coast Guard, the Fall season coincides with promotion season. This is where hundreds of officers find out that they have been selected for promotion to the next rank. It is a joyous time filled with congratulations and celebratory social media posts. A lot of individuals have worked hard, and families have sacrificed to make these promotions possible.

There is another side, however. There are also hundreds of officers that also worked hard with families that sacrificed as well, that are not celebrating. Instead of seeing their name on the promotion list, they will receive a phone call from their supervisors and be notified that they were not selected. This time of celebration for many will be a time of disappointment for them. Due to organizational regulations, this notification could also mean retiring from the organization and starting a new chapter in someone's life. A lot of mixed emotions flying around in the same workspaces throughout the country. Who is calling these individuals in their dark moment?

I have recently gotten into the habit of reaching out to my mentees when they face these moments to render support. I check up on their mental state. Watching others celebrate at work and online can be quite difficult. The questions of self-worth can easily creep into their minds. We discuss what their desires are for the future. Do you want to continue to strive for promotion the following year? Are you setup for a second career?

These conversations have led to deep discussions on career goals inside the organization and personal goals outside the organization. Plans have been put together that have led to individuals being selected for promotion the following year and plans to get a jump start on the next career move. Some of the plans have not led to being selected the following year. No matter the eventual outcome, what has always been consistent is the gratitude and appreciation expressed for the effort in showing that I cared. That I was willing to invest in them at a time that they were questioning if they were worth investing in. That I helped them look at the situation differently and determine what truly matters to them. The organization only requires individuals to be notified when they were not selected for promotion. Intrusive leaders go well beyond just the notification.

Example 5: Ask and then Truly Listen

Never underestimate the power of a simple question like, *"How are you doing today?"* A simple and routine question that normally would illicit a simple and routine response, could open the door to a much needed conversation about a serious concern. In the beginning of the book, I discussed a moment in which I was not okay as I struggled to process the racial tension in 2020 as a result of the murders of George Floyd, Breonna Taylor, Ahmaud Arbery, and the public demonstrations in city streets across America. The simple question of, *"How are you doing?"* would have gotten a different response than my normal, *"I'm busy, but I'm blessed!"* My response in that moment would have indicated that I wasn't okay, that I was finding it tough to go about my normal duties and not think about if my life mattered in the same manner as others.

In the discussion about emotional intelligence as a characteristic of an intrusive leader, I discussed a situation where Vice Admiral Manson Brown asked a group of women under his command, *"How are things at the unit for women?"* That question with the specific intent to gain an understanding of a unique perspective from a specific group of people led to a concerning response. Listening to that concerning response and making the decision to act, led to the discovery of sexual harassment at the unit. Vice Admiral Brown asked a question and then he listened.

In early 2022, I had the amazing opportunity to discuss Intrusive Leadership with the Sergeant Major of the Army Michael Grinston. In this position, he is the most senior enlisted member of the US Army and serves as an advisor to the Chief of Staff of the US Army. This is a position of immense responsibility. During our discussion, Sergeant Major Grinston told me a story where he asked a fellow soldier, *"How are you doing?"* as he was leaving the dining facility.[32] The response that he got back was not the normal cheerful and positive response that he was used to hearing. Recognizing the difference, he

---

32. SMA Michael Grinston, Interview with Sergeant Major of the Army, January 18, 2022.

told the soldier that he would like to meet with her once he got back from his security patrol. At the time he was deployed to a war zone and had a mission to complete but he also knew he had another mission to be there for his soldiers. When he returned from his patrol that evening, the soldier found him, and they talked. The situation that was concerning the soldier had nothing to do with the deployment or mission operations. She was going through a divorce and was struggling with all the emotions involved with what was currently happening at home. This was the first time she opened up to anyone at the job about what was happening but when asked a simple question, she took a chance and became vulnerable. Sergeant Major had enough emotional intelligence and maturity to allow her a safe space to be vulnerable…vulnerable in the middle of a war zone.

This moment of Intrusive Leadership was taking place while on a deployment while conducting defense operations in the Middle East. I asked Sergeant Major Grinston if he solved her problem, and his response came back quick. *"Oh no! She didn't need me to solve her problem. She just needed me to listen and let her know she was going to be okay."* A profound statement that leaders, and husbands, across the world need to understand. Sometimes listening is the only thing that is needed to make someone feel valued and supported. Don't let the fear of not having the answer or the solution stop you from showing your employees and officemates that you see them. You want them to feel heard…ask them a simple question and then truly listen.

Example 6: Upward Intrusive Leadership

Throughout my career, the discussion on leadership mostly involved how a person leads their subordinates. Often, the discussion included how people can lead upward and impact decisions made by individuals senior to them. *"Leading Up"* has been a consistent theme. That theme is relevant with Intrusive Leadership.

As I mentioned earlier, I served as the Military Aide to the Commandant of the US Coast Guard. I don't think I conveyed the risk associated with this assignment. This was the type of job that would either catapult my career to a higher level or bring my career to a quick and disappointing conclusion. Every day on the job involved moments where failure could mean being fired. Having a bad day wasn't an option in a highly visible position such as using Intrusive Leadership up the chain.26 Rear Admiral Thomas is a trailblazer for women leaders in the military, a staunch advocate for diversity, equity, and inclusion, and has personally invested time to support me throughout my career. I am not the only one either. Rear Admiral Thomas has touched the lives of hundreds of people throughout her career, making them better service members, better leaders, and simply better people. She spent so much time and energy investing into others that she often times neglected to invest in herself. This neglect manifested itself into a weight problem that almost derailed Rear Admiral Thomas' career. [33]

At a critical point in time, when her Commanding Officer communicated that he would hold her accountable for adhering to the weight standards, Command Master Chief Buck Ward provided Intrusive Leadership. Master Chief Ward called this. I was excited to be selected for the position. The complete and honest truth was that I was both excited and nervous.

As I prepared to begin this assignment, I met Special Agent Miguel Rivera. He was assigned as the Lead Agent for the Commandant's Protective Service Detail for Coast Guard Investigative Service. His responsibility was to keep the Commandant safe. The same role the US Secret Service has protecting the President is the same role Special Agent Rivera had protecting the Commandant. We were going to be working hand and hand to support *"The Boss"* and we had to work as a team to be successful.

To make sure we got started on the right foot, Special Agent Rivera scheduled a meeting. He invited me out to dinner at a local restaurant with my wife. That's right…we didn't meet in an office at the job or grab lunch with just the two of us. Special Agent Rivera made sure that I knew Miguel, who was married to a remarkable woman named Lorna. He wanted to know Marcus, who somehow won the lottery and married another remarkable woman named Angel. We discussed our children. We discussed our marriage stories. We discussed a lot of things that had nothing to do with the job. When the night came to a close, Miguel told my wife that he and his team had my back, and they weren't going to let me fail. He was true to his word, and I will forever be grateful that he did.

---

33. RADM Cari Thomas, Interview with the Coast Guard Mutual Assistance CEO, January 20, 2022.

Rear Admiral Cari Thomas, US Coast Guard retired, provided me with another example of a person then Commander Cari Thomas into his office and did something that would change her forever. He closed the door once she was in the office and looked her right in the eye and said, *"Damn it Commander, get your stuff together! The Coast Guard needs you!"* Then he went over to the clothes closet and took one of the Cutterman's pin (this is a uniform device that indicated that he had served successfully underway on Coast Guard cutters for at least 5 total years.) off one of his uniforms and handed it to her. He gave it to her and said, *"Keep this and when you look at it, let it remind you that I am right there kicking you in the butt and cheering you on."* That is all it took.

Commander Thomas got her stuff together and continued to ascend within the organization. As she rose, she never forgot to reach back and bring others up. When she heard of someone under her supervision facing administrative action because of weight, she would personally workout with them. She shared her own story and stood next to them during their time of struggle. Rear Admiral Thomas was changed by Intrusive Leadership. Intrusive Leadership provided by someone that she outranked, and she changed others with Intrusive Leadership as well.

Example 8: Addressing The Root Cause

When Rear Admiral Brian Luther, US Navy (retired), was the commanding officer of the USS George H. W. Bush (CVN-77), he had a systematic approach to reduce conduct violations by crew members. (RADM Brian Luther interview, May 16, 2024) Throughout his career leading thousands of men and women, he noticed a correlation between a lack of sound financial management and conduct problems. When a person is struggling managing their finances, they tend to have marital problems or problems with alcohol. The stress in one area shows up by unraveling in another area. RADM Luther decided to be an intrusive leader in order to reduce the amount of conduct violations.

RADM Luther decided to focus on the financial management of his personnel. He realized that some of the junior personnel had never had a paid job before. They had never been in a position to even need to budget their money. So, he started to engage with them about their budgets, their financial goals, and investments. He would provide them material on savings plans and other financial management material. As the financial literacy of his people increased, he noticed positive performance picked up, their desire for professional development enhanced, and a decrease in adverse discipline.

Through intrusive involvement in the personal finances of his people, he positively impacted their lives at home. His focus on something that was solely for the benefit of the military member who displayed a genuine care for his people, and his people appreciated it. His people worked harder, they were more vested in their military careers, and they wanted their unit to succeed. They also stayed out of trouble more. RADM Luther noticed that his counterparts on other aircraft carriers that didn't lead in an intrusive manner had to deal with more disciplinary issues than he did on his carrier. They didn't focus on "why" a person made a poor decision, so they had to spend more time dealing with more poor decisions.

Did RADM Luther's people appreciate his leadership? You bet they did. Wives expressed their appreciation multiple times for the impact he had on their husbands. The impact he had on their family. He had a number of junior officers who served under his leadership email him to let him know that they were following his leadership example. He even had one sailor's mother thank him because her son was now planning for his future versus planning the next upgrade modification to their car.

Why was RADM Luther focused on the financial management of his people? Well, it is simple. He was taken advantage of earlier in his career and made some poor financial decisions. He didn't want others to go through the same thing, so he decided to do something about it. He decided to be the leader for others that he needed and didn't get. He wanted them to succeed where he failed. So, he decided to be intrusive.

Missed Opportunities

I will be honest when I tell you that I questioned if I should put this section in the book. A section dedicated to missed opportunities. An example of when someone needed Intrusive Leadership and didn't get it. I decided to include it because I felt that there was value in letting you know how easily these opportunities to truly connect with an employee can be missed. Or provide some insight into the reality that there are some leaders out there who simply choose not to lead the whole person.

Example 1: Ignoring Significant Life Events

This past year I sat down with a mentee that I had lost contact with but just so happened to be traveling through the city he was currently living in. We had dinner and he opened up to me about a situation that he encountered at work that would prove to be detrimental to his career. He went on to explain that one evening he was home alone, and he had some alcoholic drinks. He had too many drinks and slept through his alarm clock or maybe didn't set his alarm clock at all. He was late to work the next day and because of this perceived infraction he received an alcohol incident. The documented incident being a part of his professional record could easily be detrimental to a young military officer's career. He didn't drive drunk or get arrested for disorderly conduct. He was late for work and his supervisors chose to hold him accountable in a manner that would end his career.

Why did he drink so much at home alone? Well, he had just signed the final divorce papers and ended his marriage. He was devastated that this was happening to him at such a young age. He didn't know how to deal with these emotions and he made a poor decision. When he told me the reason, my immediate question was "Did your supervisor know what you were going through?" He had told his supervisor months before that he and his wife were beginning the process of going through a divorce and his supervisor never said a word about it again. Never checked in on how he was doing. Never offered any support. Just left him alone to deal with his personal issues on his own. All the while still going about business as usual.

My mentee made a mistake, but in my opinion, it was a reasonable mistake. He was late to work because he drank too much as a result of being depressed about a life-changing event. Heartbroken and unsupported. He made the mistake but the person that should have been held accountable was his supervisor. His supervisor should have been following this personal situation closely. His supervisor should have been checking in. He should have offered some support. His supervisor instead chose to be a manager, instead of a leader. He could have been the difference, but instead, the organization is going to lose a good person. His supervisor missed the opportunity to be THAT leader, and I hope you don't.

Example 2: Not Counseling The Whole Person

One of my National War College (Class of 2022, Committee 10, Cherry Parrots, President's Cup Winners) classmates explained a counseling session with a mentor that didn't go well and the impact it had on her. She told me that she was at a point in her life where she desired to start a family and was looking to discuss how to manage family desires with her career. She requested to meet with a previous supervisor who she enjoyed working for. A supervisor that she thought would address her needs and concerns. Unfortunately, she got something different. When she expressed her desire to request an assignment that would allow her time and flexibility to begin a family, the perceived negative impact of her career became the sole focus of the conversation. The supervisor explained that her assignment preference would throw her off the track that he felt she should stay on. A track that would lead to the same career success that he had obtained. He wanted her to desire to be like him. He made the counseling session more about him than he did about her personal desires.

She left that counseling session disappointed, discouraged, and with the understanding that she had to choose family or career. She couldn't do both. As I listened to this high-performing, remarkably intelligent woman explain to me how that counseling session with a non-intrusive leader created her intentions to separate from the military, I wondered how different that counseling session could have been. I can only speculate what type of impact it would have had if my classmate received comprehensive mentoring in that moment. The type of mentoring that factored in her desires as a mother as well as her desires as a military officer. Who knows how much further she would have gone in her organization or how long her organization would have benefited from her passion.

Example 3: A Missed Opportunity to Save a Life

This past year I had the pleasure of getting to know Petty Officer Michelle Oravitz. She is currently serving in our military and has a bright future ahead of her. Unfortunately, that bright future is connected to a dark situation in her past.[34]

While Petty Officer Oravitz was assigned to a ship, she worked with another young petty officer. This petty officer was a little rough around the edges and recently got reprimanded for a conduct violation. Other people on the ship wrote him off as problematic but Petty Officer Oravitz decided not to give up on him. She gave him a chance to make a course correction and change his reputation.

The progress made was short-lived. Pretty soon after the troubled service member began to have some emotional outbursts and clashed with a few of the other unit members. Petty Officer Oravitz attempted to find out what was going on but was swiftly brushed aside. She got concerned about his well-being and decided to bring her concern to the Command Enlisted Advisor. This was the person that was the designated person tasked with bringing up concerns of the crew to the unit's Commanding Officer. This was the person that Petty Officer Oravitz should have brought her concerns to so they could be addressed. Unfortunately, she was accused of having an ulterior motive when she expressed concern about her shipmate's well-being. She decided not to stop there and went straight to the Commanding Officer. This was a pretty gutsy move on her part but unfortunately, she was accused again of only trying to make herself look better by bringing someone else down.

A few weeks later, Petty Officer Oravitz's concerns were proven right when the troubled service member took his own life. To make matters worse, the Command focused on covering their tracks versus owning their mistake. They took no accountability, displayed no humility, and they did very little to show care for the life lost.

Who knows what different outcome could have occurred if someone listened to Petty Officer Oravitz. I can still hear the frustration in her voice when she tells the story. She knows that the outcome could have remained the same,

---

34. (Michelle Oravitz interview, July 30, 2024)

but never getting the chance still makes her upset. The only thing positive that came out of this situation is the promise that she made to always do what she can to care for her people. She promised to be an Intrusive Leader.

The Geezus Technique

The last example is not an example, it is a technique. Without question, the best thing about attending the National War College (Class of 2022, Committee 10, Cherry Parrots, President's Cup Winners) was the amazing people that I was fortunate to meet and interact with daily. My classmates were absolutely phenomenal, and I will forever be grateful for the time I got to spend with them. A few classmates stand out to me and one of those classmates was Lieutenant Colonel Aaron "Geezus" Cavazos, US Air Force. There are two distinct reasons why he stood out to me. First reason is the consistent sacrifice he made throughout the whole year to ensure our class had what we needed to excel. I truly believe that our experience studying national strategic policy would have been vastly different if it wasn't for his efforts. At the end of the year, Lieutenant Colonel Cavazos and Colonel Joshua Berry, US Army, were recognized by our class.

That recognition should continue into eternity. These two leaders had the responsibility of managing Teddy's bar in Roosevelt Hall. With everything that needed to be done to earn our Master's degrees that year, and still facing the COVID-19 pandemic, Teddy's remained fully stocked with refreshing beverages, and open for students to gather and engage in intriguing conversations and debates. An amazing feat worthy of recognition.

The second reason Lieutenant Colonel Cavazos stands out to me is because of his systematic approach to Intrusive Leadership. It is a technique that had to be included in this section once I heard about it. [35]

Let's first look at how he became a leader that values the impact of Intrusive Leadership. Just like many others, Intrusive Leadership had an impact on him. Lieutenant Colonel Cavazos has a personal life story that needs to be told in a book or movie. He came from humble beginnings, faced a lot of adversity, and saw a lot of tragic incidents to become an F-35 pilot and

---

35. Lt Col Aaron Cavasos, June 13, 2022.

Commanding Officer of an Air Force operational squadron. I hope he tells his story one day because a lot of people would be blessed to hear it. It is a true American success story.

Early in his career, his mother unexpectedly collapsed near her home in Texas while he was assisting the A-10 aerial demonstration team in New York. Hours later, the hospital in Texas told then Captain Cavazos that his mother was brain dead, would never recover, and that he needed to make the call to take his mother off life support that day. A few hours later, he made the call from his New York hotel room to remove his mother from life support. Captain Cavazos then began preparations with his unit to figure out how to get to Texas and somehow get the A-10 he brought to New York back to Georgia in the process.

Fortunately for him, he had a relationship with an intrusive leader that understood that he needed to support Aaron, the only child who just lost his mother. He needed to be more than the supervisor of Captain Cavazos, a military officer that needs to go on emergency leave of absence and his duties needed to be covered by other members of the squadron. This intrusive leader talked to Aaron for as long as Aaron needed him to. He let Aaron be emotional. He helped Aaron talk through his emotions. He was exactly what Aaron needed and these intrusive conversations continue today. He has become sort of a father figure to Aaron.

The way this supervisor cared for Aaron, changed how Lieutenant Colonel Cavazos led other members of the Air Force from that point on. He felt supported and valued when he needed it most and he wanted others to experience that same feeling. So this military pilot, who normally finds it difficult to open up to people, developed a technique to be an intrusive leader and change lives.

Using an excel spreadsheet, Lieutenant Colonel Cavazos, kept track of every member of his command and filled out his spreadsheet to help him remember important things about them. He kept track of required training courses, qualifications obtained, and other relevant professional information. He kept track of members' important career milestones, life goals vocalized by the member, and used this sheet to plan out how to achieve these goals for each person. He also kept track of personal information about their upbringing, their families, hobbies, and personal desires.

From the moment they entered the unit, they were placed on the spreadsheet and Lieutenant Colonel intentionally looked for ways to connect with them. It wasn't just that he had a spreadsheet of his members. A lot of commanders have one. It was how he used it. Lieutenant Colonel Cavazos forced himself to look at this sheet every Monday and every Friday at a minimum. Mondays were for going line by line, member by member, to see what he could do that week to help each member, or their families achieve their goals. Fridays were the same but were also used to schedule next week's Intrusive Leadership conversations with 2-3 members.

This technique ensured that 3-4 Intrusive Leadership conversations were held with each member once a year. He was also transparent about his upbringing and the adversity he faced in his life. He worked to build trust. This allowed him to be the leader they needed him to be for them individually. They looked at him differently because he treated them differently. They wanted him to be successful as a Squadron Commander because he placed their success as Airmen and as people as a priority. Lieutenant Colonel Cavazos looked at the spreadsheet weekly and scheduled meetings with members that he hadn't met with in a while. He made the concerted effort to take specific actions to do more than just prepare for a deployment or complete a mission. He connected with his people, that allowed him to address family situations, place members in positions that better aligned with their desires, helped them achieve career success and personal goals.

How did his members show that they appreciated Lieutenant Colonel Cavazos' use of Intrusive Leadership? They came to his defense when he needed it. While he was serving as Squadron Commander, a junior Airman was transferred to his squadron and given an opportunity. This Airman had a troubled past in their short time in the Air Force and this was their last chance. Lieutenant Colonel Cavazos' squadron had a reputation of turning around the careers of some troubled Airmen in the past so the hope was that this latest transfer would have the same result. It is no surprise to me that airmen who struggled in other situations, flourished when they were in a situation that included an intrusive leader. Not everyone can be saved unfortunately. Within days of being re-assigned, this airman took some actions that got her into some legal trouble.

Lieutenant Colonel Cavazos held her accountable, and she claimed discrimination. When the members of the squadron found out that Lieutenant Colonel Cavazos was under investigation for workplace discrimination against an African American female Airman, the other African American Airmen in the unit lined up to make statements in his defense. The investigation was quickly concluded, and Lieutenant Colonel Cavazos was cleared of any wrongdoing. Intrusive Leadership had enabled him to be there for his members, and for his members to be there for him. They worked together as a tight knit unit serving our country and we are all better for it.

Now, one thing that I do want to discuss is what an intrusive leader should not do if they encounter an employee that is truly private. Not everyone is going to open up to you despite your intentions and your trust building actions. Some people just want to keep their personal lives private and that is okay.

When I conducted that quick survey for my research paper that I mentioned earlier, 19% of the respondents indicated that they would not share a personal issue with their supervisor at work even if it was impacting their performance. For various reasons, there are certain people that would rather remain silent. Intrusive leaders can't force their way in or demand it. That simply will not work, nor is it a requirement for an intrusive leader to be involved in the personal life of every member. What is a requirement is that an intrusive leader communicate to every member that they are available to them to discuss personal matters if the member allows it. The member has to let you in. It is the responsibility of the supervisor to earn this level of trust with their employees. Simply having the leadership position doesn't automatically mean people will trust you. It has to be earned and the supervisor is the one that has to earn it.

I often use this analogy when discussing this point. Intrusive Leadership is knocking on the door of someone's house and asking to come in. If they let you in, then great. If they don't, then we can talk all day on the porch.

We can talk week after week on the porch until they feel comfortable letting me into the living room. As long as they know that I am there to support them to the fullest extent and that I value them as person, not just as an employee.

Trust might build up through witnessing Intrusive Leadership in action. A private person can see their co-workers being supported in a manner that they actually desire and didn't even know it. They might witness a co-worker being supported as they manage the healthcare for an ailing parent. They might witness another co-worker being supported as they deal with a divorce, or a special needs child. They might witness the connection between their co-workers and supervisor, the joy that it brings, and decide that they want that too. They might take a chance and step out of their comfort zone. They might not. Revelation 3:20, says *"Here I am! I stand at the door and knock. If anyone hears my voice and opens the door, I will come in and eat with that person, and they with me."* If the Lord can knock and wait to be let in, then surely so can we!

PART V

Impacts of Intrusive Leadership

The impacts of Intrusive Leadership are the reason why I have become a champion of this leadership style. It is not because it is the easiest or the simplest leadership style. It is definitely not because it allows someone to stay in their well-defined comfort zone. It is solely because of the immeasurable impact Intrusive Leadership has had on me and so many others. Intrusive Leadership changes lives, propels careers, and it also saves lives.

Many leadership experts will tell you how important it is to motivate and inspire employees and get them to buy-in to why their job matters. In my opinion, before leaders get employees to understand the why, leaders must understand the who. Who are your employees? What will actually inspire them? What are their motivations? What issues are they facing? Intrusive Leadership opens the door to truly knowing who your employees are and what will get them to reach their full potential. Let's discuss these impacts in more detail.

Employee Satisfaction

The impact that employee satisfaction can have has been well documented. An employee that is happy at the job is an employee that works harder, stays longer, and is more committed to the success of the team or company. That happiness is impacted by the relationships that an employee has at work and specifically their supervisor or boss.

Craig Groeschel wrote a leadership book for churches and in this book, he describes the connection between employee satisfaction and workplace friendships. In the book he discussed studies that show that employee satisfaction is boosted by close workplace friendships by almost 50%, and that people that are close to their boss are two and half times more likely to be satisfied on the job.[36]

---

36. Craig Groeschel, Lead Like It Matters: 7 Leadership Principles for a Church That Lasts (Zondervan Books, 2022), 70.

Intrusive Leadership creates those type of impactful close relationships with employees. Intrusive Leadership leads to more than an employee being happy while at work, it leads to happiness because the employee is at work. Happy because they feel cared for and supported at work. Happy because they feel emotionally safe at work. Happy because of the relationships that have formulated at work and how those relationships have impacted them. That type of happiness built by Intrusive Leadership, is simply built differently and built to last. Intrusive Leadership leads to employees working harder and increasing their performance level. Employee satisfaction directly impacts commitment to organizational success, workplace unity, and retention.

Commitment to Organizational Success

When I explained my definition of Intrusive Leadership to a Coast Guard veteran, friend, and currently the Director of the Safety and Environmental Enforcement Bureau, he quickly recalled when he was impacted by Intrusive Leadership. Director Kevin Sligh went on to explain working for a supervisor who came to him on an early Friday morning and gave him some interesting task direction.[37] After a quick discussion between the two of them about things that were happening with the assigned work list, his supervisor turned to him and said, "Okay, get what you can get done this morning, but leave at noon and go take your wife to lunch or a movie." Confused and not quite sure of the validity of these taskings, Director Sligh said he had to confirm this with some other office mates. They had received the same taskings. "So, at noon, I went home. After that day, there was nothing my supervisor could ask of me that I wouldn't do." I had to explain my definition of Intrusive Leadership to Director Sligh when our conversation started, and then he instantly explained how Intrusive Leadership got him to work harder.

Retention

Intrusive Leadership leads to the retention of quality talent. One of the first interviews that I did when I was researching Intrusive Leadership was with Master Chief Jason Vanderhaden, who was serving as the Master Chief Petty Officer of the Coast Guard at the time.30 Having read my 2020 article, Master Chief Vanderhaden was very familiar with my message on how Intrusive

---

37. Kevin Sligh, Interview with Director of the Safety and Environmental Enforcement Bureau, May 14, 2022.
38. MCPOCG Jason Vanderhaden, Interview with Master Chief Petty Officer of the Coast Guard, November 29, 2021.

Leadership was critical in addressing racial tension.

During the interview he also explained how Intrusive Leadership impacted him in another fashion. While stationed on an inland river buoy tender, Master Chief was working as the culinary specialist onboard. He was responsible for buying the food and preparing the meals for the entire crew. Anyone who has spent time serving on the high seas knows the impact that a good meal can have on the morale of a crew.

The Commanding Officer of the buoy tender demanded more from the Master Chief than just good meals. He would routinely wake up Master Chief and get the buoy tender underway and moving towards the first buoy that needed to be serviced that day. Just the Master Chief and the Commanding Officer working together while everyone else was sleeping. Master Chief recalled not really enjoying this extra work at first, but something started to change. In addition to getting the buoy tender underway, the Commanding Officer and Master Chief also spent time on the bridge talking about life. These conversations began to be something that Master Chief looked forward to.

At a time when he was considering leaving the service, a leader that demanded more of him, got him to work harder and stay in the organization through Intrusive Leadership. "If it wasn't for that Commanding Officer and those conversations, I don't think I would have stayed in much longer and I definitely wouldn't have become the Master Chief Petty Officer of the Coast Guard." he stated. Time and time again, I heard that statement was for senior leaders. Their amazing talent was retained by organizations as a direct result of Intrusive Leadership.

I mentioned Chief Jason Kuzik earlier in the section about transparency. Before he told me the story about his openness with his members, he explained to me why he led in an intrusive manner. He told me about Chief Warrant Officer Steve Cabral, U.S. Coast Guard (retired), and how his leadership impacted him. Chief Kuzik told me how he had encountered toxic leaders while previously serving in the U.S. Army and the beginning of his U.S. Coast Guard career. Leaders that made him feel insignificant and treated him in a demeaning manner. Leaders that made him leave the Army and was about to make him leave the Coast Guard as well. That all changed when he encountered CWO Cabral.

CWO Cabral's leadership philosophy was to establish positive personal relationships with his people and empower them to build up their confidence and create ownership in mission success.[39] He treated people the way he wanted to be treated because he, himself, encountered a toxic leader earlier in his career. He didn't want his people to feel the way he felt, so he decided to lead differently.

He was very intentional in treating everyone with respect and made sure he knew them beyond the job. He met with wives and opened the unit up to family members. He held an open house for the local community to build a better relationship with the constituents in the surrounding area. He took these actions with the objective of earning trust with his members, their families, and the local community. This different approach to leadership was the difference for Chief Kuzik and others.

When CWO Cabral arrived at the Coast Guard station where Chief Kuzik was assigned, he noticed that the crew seemed beat-down and had low morale. It became obvious that the unit was facing some serious leadership challenges and CWO Cabral was sent to be the fix. The leadership style of the previous Commanding Officer wasn't toxic, but it wasn't engaging either. It wasn't personable like CWO Cabral's leadership style. His leadership style was known at this point in his career, so he was hand-selected as a result. Thankfully for Chief Kuzik that was the right selection.

It was also the right choice for Lieutenant Commander Johnston Ariail as well. LCDR Ariail was serving as a Second-Class Petty Officer when he was assigned to the small boat station. At the time CWO Cabral arrived, Ariail was planning to get out of the Coast Guard like many other members in the unit. Shortly after CWO Cabral took over, LCDR Ariail was placed in a leadership position and given a task. The task was to review previous search and rescue missions and start writing the crew up for awards. This task of recognizing the crew and making sure they felt valued had a profound impact on LCDR Ariail. CWO Cabral's personal engagement with everyone and wanting to know about their families and life outside of work also had a profound impact. LCDR Ariail decided to stay in the organization because he wanted to make people feel the way CWO Cabral made him feel. When I asked him if CWO Cabral was the reason he was still in the U.S. Coast Guard

39. (CWO Steve Cabral interview, May 28, 2024)
40. (Johnston Ariail interview, June 10, 2024)

so many years later, his response was the same response that Chief Kuzik had when I asked him the same question. "He is absolutely the reason I stayed in!"

Any human resource professional will tell you how retaining quality talent is a priority for organizational success and increased retention directly impacts the bottom line. Millions of dollars were spent every year by companies for hiring individuals and training them. This cost can be offset by retaining talent. Any discussion about how to retain quality talent must include how that talent is led, and they must be led with Intrusive Leadership.

Workplace Unity

Intrusive Leadership leads to increased workplace unity. While speaking to a group of supervisors for the Fire and Emergency Management Service for the District of Columbia, we discussed the debate between Black Lives Matter and All Lives Matter. This debate was commonplace in America during the summer of 2020, but the debate should have never been. It should have been a discussion with the point of gaining understanding and ending with support. I explained to the supervisors during this training session that too many debates about Black Lives Matter versus All Lives Matter dealt with trying to prove the validity of statements or justify certain actions. Too many assumptions about the purpose behind rallies and protests. That shouldn't have been the focus. The focus should have been on understanding why a certain group of people felt the need to say that their life mattered.

I asked this group of supervisors that had engaged in or witnessed these debates in fire houses across the District of Columbia if any of them ever communicated to their African American members that their life mattered to them. "Did you ever look at them and say, "I can't change what is happening on the street, in the news, or online but I can see your pain and concern. I can't say that I know exactly what it is like to be in your shoes and experience life as you do. But I need you to know that here, in this fire house, amongst these men and women, your life matters to us. Your life as an African American matters to us." That was never stated. Imagine jumping on a fire truck because the fire alarm went off, driving down the street towards a blazing fire, willing to sacrifice your life to save someone else, and questioning if the other people on the truck with you values your own life.

Whether it is a fire truck, rescue helicopter, or corporate boardroom, why would we not want everyone to feel that sort of support and care? The Asian community faced some challenges with domestic terrorism during the COVID-19 pandemic. Did we reach out to our Asian American coworkers and place our arms around them? Supervisors understand how important it is to be unified around the operational mission or corporate goals. Supervisors that understand the impact of Intrusive Leadership know how important it is to be unified around the support and care for every employee as well.

Companies and organizations that place a priority on employee satisfaction will win in this global competitive environment. Intrusive Leadership plays a role in creating a happy workplace, increasing employee productivity, retention, and workplace unity.

During this past year, I was fortunate enough to be a panelist for a leadership discussion held during a training conference at the Defense Information System Agency. I think I did a good job, but anothr panelist stole the show! This phenomenal leader provided a leadership example that had to be included in this edition, so I tracked her down and made her tell me the story again. [41]

Ms. Tara Walker, who asked that her identity be concealed to protect her team member's privacy, had just completed a leadership development program and was assigned to her first true leadership assignment within the Federal government. Before meeting her team, she had some meetings with the person she was replacing and her new supervisors. During both meetings, a troubled employee was discussed. An employee that was disengaged, disinterested, and labeled as a problem. Ms. Walker had been warned but she decided to give everyone a fair chance at a first impression.

In the first meeting with Jane, Ms. Walker asked about her role with the team. The disengaged and disinterested label was confirmed. Jane only responded with facial expressions and when she did decide to say something she indicated that she was assigned to a team that she didn't request. She indicated that she wanted to work a budgetary job instead. The first meeting ended on a less than positive note, but Ms. Elsey Walker wasn't willing to label Jane as a problem just yet. Shortly after the meeting, Jane requested to move her to a desk away from the team and into a corner of the office. She

---

41. (Tierra Elsey Interview, June 27, 2024)

isolated herself but Ms. Walker had a plan. Her plan was to allow Jane to isolate herself from the team while she took intentional actions to get closer to her.

Ms. Walker recalled that Jane said that she had an interest in budgetary work and also discovered that she also liked administrative. Ms. Walker decided to give Jane some tasks that aligned with her desires, and Jane took ownership of them. One of the tasks was to organize the teams email inbox which at the time was in disarray and creating an ineffiient process. Jane cleaned it up and smoothed out the email monitoring and response process. It quickly became clear that Jane had the ability to positively contribute to the team.

In addition to Ms.Walker making adjustments with work tasks, she also made an intent to engage with Jane on a personal level as well. They bonded over their mutual love for dogs and their relationship blossomed from there. Jane started to become more involved with the team, and the team started to engage with Jane more consistently. Jane sought feedback from Ms. Walker and positively responded to criticism when it was provided.

It took some time, but Jane began to be a positive light in the office. She enjoyed baking and she would occasionally bring in baked treats for the office. She improved her communications skills by taking Ms.Walker's advice and began to go above and beyond. She came into the office to handle critical administrative tasks during the COVID pandemic without being prompted. She truly became an integral part of the team's success.

To Ms.Walker's surprise, one day Jane came to her with a simple request. She requested to move her to a desk that was surrounded by the rest team and no longer be separated from her teammates. The unity in workplace was positively impacted. Jane was never a problem employee. She just needed the right leader to connect with her. Ms. Walker was that leader for Jane, and she remains in contact with Jane today even though Ms. Walker works for another agency now.

Diversity, Equity, and Inclusion (DEI)

It has been well-researched and documented that diverse teams outperform non-diverse teams. That debate has been settled. So how does Intrusive Leadership impact DEI? Let me first explain how I conceptualize DEI. I call it the "party example" and it goes like this:

Diversity is everyone being invited to the party. Every person no matter race, religion, gender, or sexual orientation is invited to the party. If you want to go to the party, then you can go to the party. Nobody is not invited or prevented from attending the party. Equity is everyone having the same access to the dance floor. Everyone has the same opportunity to go on the dance floor and dance. Nobody has to do anything extra to get to the dance floor. If they want to dance, they can dance. Inclusion is the type of music being played. Even though, everyone has the same invitation to the party and has the same access to the dance, how a person feels or enjoys being at the party will depend on the music being played. If the music that they enjoy is being played, then they are going to enjoy being at the party. They are going to dance the night away and tell their friends about how great the party was. The people that don't like hip hop are going to have a completely different experience at the same party if hip hop is the only music that is played. They might show up just to appear like that they are a part of the group, but they aren't going to be truly motivated. They might even dance a little just to make sure they are seen, instead of dancing like nobody is looking. An inclusive party is one that everyone's favorite music is played throughout the evening. A party where the DJ makes sure everyone can be their authentic self and enjoy themselves.

Intrusive Leadership allows a supervisor to know what music their employees need to hear to be their authentic self at work. Intrusive Leadership allows a supervisor to create a work environment where every employee is excited to come to work and feel like their desires and needs are valued. Too many supervisors have a leadership style that they like and are comfortable with. It is typically a leadership style that worked best for them. They take that leadership style and peanut butter spread it on every employee. Those employees that respond to that specific leadership style rise to the top. The rest of the employees are left to operate under a leadership style that doesn't connect with them and still try to rise.

There are too many employees that have been rated as poor dancers because they can't dance to hip-hop, when they are amazing salsa dancers. Salsa music was never played. They spend years showing up to a party that they don't truly enjoy and acting like they like the music. A leader that understands the impacts of Intrusive Leadership, knows who likes hip-hop, salsa, or country music and uses that knowledge to get the best out of each employee.

They create a work environment where everyone is working together while dancing to their favorite music.

So, how do you define Intrusive Leadership for Diversity, Equity, Inclusion (DEI)? Here is my definition of Intrusive Leadership specifically for Diversity, Equity, Inclusion purposes:

Intrusive leadership is a leadership style that uses intentional actions to inspire, motivate, develop, and support people by caring for the whole person through a trusting relationship that addresses professional as well as personal interests, desires, and concerns.

Do you see the difference? Did you notice what changes when you consider this leadership style regarding DEI? There isn't a change. There isn't a difference. If a leader is intentional about getting to know everyone and building a trusting relationship with a deeper level connection, then that would create the inclusive environment that so many organizations aspire to create. This isn't a leadership style that you do for women or minorities, this is a leadership still that can be impactful for anyone. Managers can do a lot to make sure that a workplace is void of discrimination and harassment towards minorities and still fail to go the extra mile and make the minorities feel seen. I will speak for myself when I say that I don't want to have to hide or mask the pride that I have in being an African American at work. Not dealing with discrimination is the bare minimum, but building authentic relationships with my co-workers is highly desired.

I want my supervisors to take an interest in how my family is doing. I want them to consider how certain events might impact me differently than my peers. I'm not saying that I only want my ethnicity to be seen, I don't want racism or the "African American experience" to be the only thing that is discussed. However, I don't want it to be ignored either. Organizations and managers need to take an additional step beyond the cultural observances and affinity group programs and look to build real authentic relationships with everyone. Learning about the Tuskegee Airmen is great, but making sure I feel valued and supported as one of the few African American pilots at my unit is critical. Intrusive Leadership does that, and it can do that for everyone.

Intrusive Leadership enables people to break diversity barriers. During my research, I was fortunate enough to talk to three Browns that are trailblazers impacted by Intrusive Leadership. They have all broken diversity barriers and rose to the highest level of the military. General CQ Brown Jr. became the Chief of Staff of the Air Force and the first African American to lead a military organization as the Service Chief. Vice Admiral Manson Brown became the first African American Vice Admiral in the US Coast Guard. Rear Admiral Erroll Brown became the first African American to achieve the rank of Admiral in the US Coast Guard.

Each one of these amazing leaders were impacted by Intrusive Leadership in a manner that allowed them to remain focused on excellence and committed to their organization. That focus and commitment allowed them to achieve feats without a visual representation of someone that looked like them achieving the same feat.

Early in his career, General CQ Brown Jr. served as a military aide to the Chief of Staff of the US Air Force. Just like my assignment as the military aide to the Commandant of the US Coast Guard, this was a very rewarding high-profile assignment. A high-risk job that would catapult anyone's career if successfully completed. While in this assignment, General CQ Brown Jr. drafted his post-Air Command and Staff College recommendation to return to the flying. Typically, completing an assignment like this is a person's golden ticket to another rewarding follow-on assignment that would continue to propel their careers. The General provided his recommended comments to the Chief of Staff and waited for the signed endorsement. Assuming that the recommendation would be endorsed, he was surprised when he the paperwork directed a staff tour following school. Confused as to what this meant for him and the reason why, he sat down with a senior leader to gain insight.

During this session, the senior leader drew a simple timeline of his career moving forward. He then drew another timeline that included milestones for his children and his family. The supervisor knew the ages of the General's children so he indicated when they would be graduating high school and starting college. He listed out future assignments and promotion opportunities. The supervisor went on to explain that the next series of assignments, that offered the best opportunity for development and promotion. The senior leader knew that General CQ Brown Jr. was a very driven and talented fighter pilot.

He knew that General CQ Brown, Jr. had great potential for senior ranks in the organization and made it a point to consider family. *"If I had followed the plan I wanted, I probably would have retired as a Lt Colonel,"* the General stated.

Now, in full transparency and to ensure that I don't misrepresent the General's current position on Intrusive Leadership, I must share that he considers this as active or comprehensive mentoring. A type of mentoring that is more than mentoring on mission and career, but also on life. In discussing this situation the General asked me "What is the difference between Intrusive Leadership and active or comprehensive mentoring?" My answer is that I don't see a difference. I see a relationship.

This sort of mentoring that goes beyond the workplace and addresses factors that impact the whole person is an action of Intrusive Leadership. This is exactly how intrusive leaders engage with their employees. They know them, they deal with them as people, and show them how much they are valued beyond their job description. This sort of intentional action led to General CQ Brown Jr. reaching the highest levels in the US Air Force and breaking diversity glass ceilings along the way.

Vice Admiral Manson Brown was at a crossroads in his career when he was assigned to a unit in Miami, Florida. He had finished engineering graduate school, built a marketable resume, and had a young family. He was considering leaving the US Coast Guard and taking his talents to the private sector. Then he met Commander Greg Magee, a supervisor that ended up changing his life. Vice Admiral Brown explained to me that when he was supervised by Commander Magee, it was the first time that he felt like he was a part of the "inner" circle at work. By engaging with him on a routine basis, Commander Magee got to know Vice Admiral Brown at a deeper level. He knew about his family and his children. He engaged with him in a way that made him feel special. Vice Admiral Brown wasn't alone in that feeling. He explained that Commander McGee made everyone feel that way. He was consistent in taking intentional actions to connect with everyone. Before I was assigned to Commander Magee's unit, I was considering leaving the organization. After I experienced his leadership style, the Coast Guard was going to have to kick me out. I was all in after that and I wanted everyone to feel the same way Commander Magee made me feel.

Well, the organization didn't kick him out and Vice Admiral Manson Brown became first and currently the only African American to achieve that rank in the US Coast Guard. At a time when he was considering leaving, Vice Admiral Brown was led by an intrusive leader, and he chose to stay. He decided to stay and become something that he hadn't seen someone that looked like him become.

Rear Admiral Erroll Brown has a slightly different story involving Intrusive Leadership.31 When Rear Admiral Brown was a Lieutenant and assigned to US Coast Guard Headquarters in Washington, DC, he was performing well. He was doing his job, completing tasks, and staying out of trouble. Everything was fine.

One day, out of the blue, he received a phone call. A phone call from Commander Merle Smith, the first African American to graduate the US Coast Guard Academy. At this point in time, Lieutenant Brown didn't have a relationship with Commander Smith, but he knew exactly who he was. Lieutenant Brown graduated the US Coast Guard Academy just six years after Commander Smith and was serving at a time that every African American officer at least knew of each other. Commander Smith had earned the respect and trust of Lieutenant Brown because of what he endured, what he represented, and the excellent manner in which he served. The phone conversation that day was short, but it changed the life of Lieutenant Brown. What Commander Smith said was simple, direct, and it was intrusive. He said "Lieutenant Brown, I have been following you throughout your career and I know you can do better." He was right and Lieutenant Brown knew it. Lieutenant Brown was doing his job, but he was not focused on being excellent. He wasn't getting in any trouble, but he also wasn't trying to be the very best he could be. He knew he could be doing better and being confronted with that reality by Commander Smith struck him at his core.

That strike to the core was exactly what he needed to inspire him to be excellent and perform at the highest level. Lieutenant Brown went on to be Rear Admiral Brown, the first African American to be promoted to Flag Officer in the US Coast Guard, and now he is known to every officer in this organization today. children. He engaged with him in a way that made him feel special. Vice Admiral Brown wasn't alone in that feeling. He explained that Commander McGee made everyone feel that way. He was consistent in taking intentional actions to connect with everyone. Before I was assigned to Commander Magee's unit, I was considering leaving the organization. After I experienced his leadership style, the Coast Guard was going to have to kick me out. I was all in after that and I wanted everyone to feel the same way Commander Magee made me feel.

Well, the organization didn't kick him out and Vice Admiral Manson Brown became first and currently the only African American to achieve that rank in the US Coast Guard. At a time when he was considering leaving, Vice Admiral Brown was led by an intrusive leader, and he chose to stay. He decided to stay and become something that he hadn't seen someone that looked like him become.

Rear Admiral Erroll Brown has a slightly different story involving Intrusive Leadership.31 When Rear Admiral Brown was a Lieutenant and assigned to US Coast Guard Headquarters in Washington, DC, he was performing well. He was doing his job, completing tasks, and staying out of trouble. Everything was fine.

One day, out of the blue, he received a phone call. A phone call from Commander Merle Smith, the first African American to graduate the US Coast Guard Academy. At this point in time, Lieutenant Brown didn't have a relationship with Commander Smith, but he knew exactly who he was. Lieutenant Brown graduated the US Coast Guard Academy just six years after Commander Smith and was serving at a time that every African American officer at least knew of each other. Commander Smith had earned the respect and trust of Lieutenant Brown because of what he endured, what he represented, and the excellent manner in which he served. The phone conversation that day was short, but it changed the life of Lieutenant Brown. What Commander Smith said was simple, direct, and it was intrusive. He said "Lieutenant Brown, I have been following you throughout your career and I know you can do better." He was right and Lieutenant Brown knew it. Lieutenant Brown was doing his job, but he was not focused on being excellent. He wasn't getting in any trouble, but he also wasn't trying to be the very best he could be. He knew he could be doing better and being confronted with that reality by Commander Smith struck him at his core.

That strike to the core was exactly what he needed to inspire him to be excellent and perform at the highest level. Lieutenant Brown went on to be Rear Admiral Brown, the first African American to be promoted to Flag Officer in the US Coast Guard, and now he is known to every officer in this organization today.

There is another aspect of Intrusive Leadership that is relevant to this diversity, equity, and inclusion section that is often not talked about. And that aspect deals with organizational trust. People enter organizations with varying levels of organizational trust as a result of their collective life experiences. Some people enter trusting that they will be fairly treated, fairly evaluated, and have a fair opportunity to exceed in an organization just like everyone else. Other people enter organizations believing that they will not and do not. Vice Admiral Manson Brown was one of those people.[42] Growing up in Washington, DC in a predominantly African American neighborhood, during the 1960s and 1970s, and watching the country wrestle with civil rights caused him to be cautious with his trust for white people. Attending the US Coast Guard Academy and being one of the few African Americans in his class provided some unfortunate experiences that further instilled that distrust.

He carried that distrust with him like a weight around his legs until Commander Magee exercised Intrusive Leadership on him. Vice Admiral Brown had experienced Intrusive Leadership from other African American officers but when he experienced it from a white officer it impacted him differently. Experiencing Intrusive Leadership from a white officer got him to drop the weight of distrust and allowed him to start having the organizational trust that is necessary for someone to achieve the career achievements he was able to achieve.

People don't spend decades performing at their absolute best and making a number of personal sacrifices to remain loyal to an organization that they don't trust values them. I can personally relate because I also carried that

42. RADM Erroll Brown, May 2, 2022.

weight of distrust. I was also one of those individuals that entered the organization with a level of organizational distrust.

I grew up as a military kid. I was moving around throughout the country and enduring unfortunate experiences. Experiences like being called racial slurs at school, not being allowed to go into the homes of certain white childhood friends in my neighborhood, and being harassed by law enforcement while doing nothing wrong.

In some respects, feeling like my abilities would always be questioned, feeling like my performance always needed to be better to receive the same recognition as my peers, and thinking that my white supervisors would always choose someone who looked like them when they were providing recommendations fueled me. Feeling like I would be mentored, while my peers would be sponsored drove me. I carried that weight around my leg with pride and was confident in the noticeable strut that weight caused. But I would be lying if I didn't admit that the distrust didn't weigh on me at times. I remember thinking that my white supervisors will always take care of the junior officers that reminded them of themselves when they were younger.

I didn't have that advantage because none of them were an African American junior officer. I expected Intrusive Leadership from other African American officers, and I got it. I can write another book about the times I was professionally and personally sowed into by African American officers who wanted me to succeed and treated me like I was one of their sons. Their Intrusive Leadership didn't force me to drop that weight around my leg. The Intrusive Leadership from white officers was unexpected and it impacted me differently. That Intrusive Leadership got me to look at the organization differently and made me feel like I could have a long and enjoyable career being my authentic self. I dropped the weight, started to trust more, and worked even harder.

Not every minority entering the organization feels the same way regarding organizational trust. General CQ Brown Jr. didn't harbor that feeling of organizational distrust when he entered the US Air Force. Yet, he still needed Intrusive Leadership to achieve organizational success. So, who is distrusting and who is not? That can only be answered through Intrusive Leadership, but no matter the trust level, Intrusive Leadership can have significant impact.

Leadership Effectiveness

Intrusive Leadership allows supervisors to be more effective as leaders. As a result of truly getting to know the people that work for you, a supervisor knows exactly what would motivate and inspire someone. They are able to speak directly to a subordinates' desires, concerns, or fears. Throughout this book I have referenced multiple stories that illuminate the type of impact an intrusive leader can have or the everlasting impression that they can leave on someone. These stories speak to the effectiveness of Intrusive Leadership. In this section, I will discuss three more ways that highlight how effective intrusive leaders can be.

Accountability

"You weren't at your best today, and the Coast Guard needed you to be." Those words pierced my professional soul as Commander Todd Lutes said them directly to me. How could he say that? Why would he talk to me like that? These were some questions that rang in my head after hearing those harsh words. The answers to those questions came quickly as well. Why would he talk to me like that? Well, because he was right. I failed to manage my crew correctly that day while they were conducting some maintenance procedures on our deployed aircraft. That mismanagement led to my crew not being available to fly when we were requested to fly for an operational law enforcement mission.

There were other courses of actions available to me that I didn't use. I should have made better decisions and I knew it. How could he say those words to me? How could he be so direct in his criticism and not create a feeling of anger inside of me? Commander Lutes was able to do so because he had won me over through his intentional actions to create a trusting relationship with me. Commander Lutes was the senior officer that played basketball with us at Air Station Miami. As a result, instead of getting angry I listened I learned the lesson he was trying to teach, and I got better. Intrusive Leadership allowed for Commander Lutes to hold me accountable when I needed it and produce the desired result. His leadership was more effective with me because he led in a manner that others hadn't.

I had a quick conversation with my friend Tarran Johnson III. after he read the first edition.[43] He currently works in the healthcare industry, and he recognized a clear difference in the ability of leaders to hold people accountable based upon their use of Intrusive Leadership. He explained to me a situation at work where there was a supervisor that struggled to hold people accountable. Not struggle in the sense of unwilling to offer criticism or provide critical feedback. Struggle in the sense of not getting people to be receptive to the criticism or feedback. This particular supervisor was not engaged with her employees and reframed from building personal connections with them. She wanted to offer critiques without making the effort to connect. She struggled mightily.

The supervisor constantly received pushback, dealt with consistent resistance to her initiatives, and was accused of discrimination multiple times. There was constant tension and significant number of hours spent on addressing complaints instead of pursuing objectives. The struggle and stress eventually forced this supervisor to leave the company.

The next supervisor came in with a different approach. She was engaging and she did build personal relationships. The need to hold people accountable was still present and necessary. However, this supervisor was able to do it and get a positive response. Tarran described her as "tough", but he also described her as someone "that people respected because she took the time to get to know people first." As a result of her intentional efforts to connect, be transparent, and humble, she was more effective in holding people accountable.

Intrusive Leadership is not a leadership style that is all about being overly friendly with people. It isn't about allowing people to perform below standards or avert rules and regulations. To the contrary, Intrusive Leadership enables leaders to motivate people to exceed standards and ensure they are adhering to the rules and regulations.

---

43. (Tarran Johnson III Interview, June 7, 2024)

Effective Communication

When I often hear people talk about effective communication, I hear how important it is to clearly communicate your vision, making eye contact, or thoroughly covering a topic. What I don't hear too often is how important it is to know who you are talking to. Who are you targeting with your communications? I think that knowing who you are communicating with is just as important as knowing what you are communicating.

"We are going to be open kimono when it comes to the survey results" is a phrase that you could easily hear in workspaces throughout the country. A phrase intended to communicate that transparency is going to be a priority. This phrase would be uttered and typically nobody would think twice about it. But how do employees of Asian descent respond to that phrase?

This is something that I would have never thought about until I talked to a co-worker of mine. He explained to me how that phrase bothers him. A kimono is a traditional Asian garment that is worn by women. So, when someone is trying to convey transparency by using that phrase, they are using the notion of opening an Asian women's dress and looking at her naked. My co-worker explained that he has tuned out senior leaders who have used that phrase because of his disgust. I can only imagine how many people have tuned me out for using that phrase. I never thought about it that way before, but it is the only way that I think about it now.

Communicating with this younger generation is something that has to be learned. Messages that resonated before don't resonate now. This younger generation thinks differently, they value different things, so communication with them has to be different as well. Too many leaders communicate in a manner that was effective for them 20 or 30 years ago. They haven't adjusted their mindset, nor have they taken the time to know how to communicate with the younger generation. They struggle to connect using the same methods that are far outdated. They become frustrated because the younger generation asks more questions and seeks a deeper level of understanding. I believe this younger generation has a lower level of organizational trust and how you communicate to them has to take that into account. The way you communicate with someone who doesn't trust you is different. Taking time to recognize each individual and speak directly to trust deficits will improve communications.

Inspiration

While I was serving as the Commanding Officer of Coast Guard Air Station Houston, Texas, I had a check-out meeting with a group of US Coast Guard Academy cadets who were assigned to the unit as part of their summer training in 2020. They spent a few weeks with us being exposed to life at an aviation operational unit and learning about the duties and responsibilities of the pilots. These four cadets desired to go to flight school and become pilots themselves so this experience at the unit was valuable to them. The check-out meeting was my opportunity to engage with them right before they left and went back to school. It was also my chance to give them some words of encouragement right before they began their senior year. During this meeting, the cadets expressed a lot of appreciation for the willingness of my crew to teach them things and expose them to operational missions. Clearly being able to go on flights was the highlight of their time with us. All four cadets were confident that becoming a Coast Guard pilot was what they wanted to do, and they were excited about the opportunity of joining the aviation community. As the meeting was coming to a close, I left them with one final thought to provide them with some motivation. I told them this:

"Outside my office you will see a row of pictures of the all the previous Commanding Officers of this unit. I'm sure you noticed that there isn't one woman on that wall. Don't let that discourage you. Don't let that give you the idea that you can't be on that wall. You have every right and opportunity to be on that wall like anyone else. Let the fact that you don't see a woman on that wall motivate you to become the first one. As a matter of fact, when you leave here today take a selfie next to that wall. When things get tough along your journey, pull that picture out and remind yourself of the goal. Remind yourself why you are working so hard and putting forth your best effort. Remind yourself that you are doing all of that and making the sacrifice to earn your place on that wall. Now go make it happen, I look forward to seeing it!"

In the Spring of 2021, the senior class at the US Coast Guard Academy received their first tour assignments. They will be assigned upon graduating, receiving their receiving their commission as an officer, and begin their professional journey. Three out of those four cadets were selected to begin flight school. Two out of those three wrote me separate emails to let me know that they routinely looked at that selfie throughout their senior year to help them get through the difficult times and remain focused. Looking at these four cadets sitting in my office and recognizing a potential way to inspire and motivate them, understanding that as women they might have different perspectives or concerns, and speaking directly to those had impact. My leadership discussion was effective because I played a type of music they could authentically dance to.

I will repeat something that I mentioned earlier in the book. Intrusive Leadership is one of many leadership styles that any leader must understand and have in their leadership tool bag. I am not saying that Intrusive Leadership is a replacement for other styles, or the only style needed. What I am saying is that Intrusive Leadership, when properly and consistently applied, can enhance the effectiveness of any leader, in any situation, with anyone. Intrusive leaders can more effectively motivate, inspire, and support people. Intrusive leaders are simply more effective.

Suicide

Intrusive Leadership can help save lives. At this current time the military, as well as American society as a whole, is battling against suicide. Suicide rates in many organizations are at a historically high level and there are numerous discussions addressing how to solve this problem. When the DoD released their annual report in October of 2022, Secretary of Defense Lloyd Austin discussed a recently established Suicide Prevention and Response Independent Review Committee to help the military attack this issue. [44] In March of 2022, U.S. News published an article discussing the spiking rate of suicide in pre-adolescent children. The US Centers for Disease Control and Prevention issued a report that revealed that close to 48,000 people committed suicide in the US in 2021.[45] That is one person every eleven

---

44. "Department of Defense Releases the Annual Report on Suicide in the Military: Calendar Year," U.S. Department of Defense, accessed January 14, 2023, https://www. defense.gov/News/Releases/ Release/Article/3193806/department-of-defense-releases-the-annual-report-on-suicide-in-the- military-cal/https%3A%2F%2Fwww.defense. gov%2FNews%2FReleases%2FRelease%2FArticl e%2F3193806%2Fdepartment-of-defense-releases-the-annual-report-on-suicide-in-the-military- cal%2F.
45. "Suicide Rate Is Spiking Upwards in Preadolescent Children," accessed January 14, 2023, https://www.usnews.com/news/health-news/articles/2022-03-15/suicide-rate-is-spiking-upwards-in- preadolescent-children.
46. Deidre McPhillips, "US Suicide Rates Rose in 2021, Reversing Two Years of Decline," CNN, September 30, 2022, https://www.cnn.com/2022/09/30/health/suicide-deaths-2021/ index.html.

minutes. Imagine that for a moment. Every eleven minutes, someone somewhere is feeling so down, void of hope, and full of despair that they see no other way to deal with the pain. Mental health is a real issue and the COVID-19 pandemic only made things worse. This was highlighted in great detail when U.S. Congress held a hearing on substance abuse and suicide risk. Companies and organizations are taking a round turn on mental health. There is widespread recognition that people are dealing with a lot of things at work and at home, and organizations need to provide mental health resources and make employees feel psychologically safe. I'm sure the company you work for has a hotline number you can call or a website that you can go to get the information you need to get help. Secretary Austin ended his official statement when the DoD released their annual report by providing the number to a hotline. The military Chaplain Corps is laser-focused on this issue in the military, as well as counselors in our school systems. This issue is getting the attention it needs but there is a role that Intrusive Leadership can play that needs to be addressed as well. [46]

How do you create an environment where someone feels psychological safe enough to come to you for help? How do you put yourself in position to have the opportunity to prevent someone from committing suicide? Intrusive Leadership is the answer.

While I was doing my research on Intrusive Leadership, interview after interview included a story where someone else stopped someone from committing suicide. Chief Master Sergeant Daniel Tester still gets choked up thinking about the day he chose to be intrusive and prevented a young Airmen from taking his life.

Colonel Kathy Barrington told me about the time she put her evening plans on hold and ended up saving the life of one of her subordinates. Rear Admiral Cari Batson told me about the moment she was approached by a Coast Guard member who told her that years before he got some help just because he heard her talk about suicide while addressing his unit.

Story after story explained how life after life was saved. Each time feeling a sense of gratitude because that life that was saved was someone's son, daughter, mother, father, sister, brother, uncle, aunt, or friend. That person got to enjoy another birthday, anniversary, or holiday season because someone in the right place choose to be intrusive.

That choice isn't always the easiest one to make. When I explained Intrusive Leadership to Major Jeff Davis, US Army, he quickly recounted a time when his IL was the difference. He explained how he was going about his Saturday with his family, but his mind was on one of his soldiers.[47] He knew that this soldier was going through a difficult time, and he wanted to check up on him. It was Saturday so he thought about just waiting until Monday and not bothering the soldier on his off-duty time. He instead chose to call him that Saturday, expressed his sole intention was to see how he was doing, and ended up stopping him from taking his life.

Major Davis said something to me that I probably will never forget when I think about the impact Intrusive Leadership can have on suicide. He said *"I would rather piss someone off by calling them on Saturday, than be pissed off doing funeral arrangements on Monday."* I won't forget that statement and I also won't forget being in position to stop one of my Coast Guard members from taking their life as well. I have vivid memories of saving lives on high seas or during Hurricane Katrina. I remember like it was yesterday introducing a woman from Texas to the aviation crew that saved her life during Tropical Storm Imelda during an award ceremony. I will also never forget watching a mentee rejoin his family after going to a dark place in his mind. I thank God that He put me in the right place to listen for a cry for help. Intrusive Leadership enabled me to hear that cry.

Over the past year, I continued to hear story after story about how Intrusive Leadership put someone in position to be the difference. During a presentation in New Orleans, LA (one of my favorite cities!), one of the attendees told the audience that he knows for a fact that the leadership style I was describing lead to two lives being saved at his previous unit. I got in contact with that person after the presentation and he connected me with Chief Kuzik. Chief Kuzik provided me with the best story that highlights the impact Intrusive Leadership can have and thankfully he allowed me to share his story. The following is his story in his own words that ended up becoming an article for the U.S. Naval Institute entitled "Intrusive Leadership Saved The Life Closest To Me."

Throughout my military career, I have always tried to make it a point to do

---

47. Major Jeff Davis, September 7, 2022.

whatever I could to make the people who worked for me know that I cared cared for them. *"People are our most precious asset!"* is a phrase that I have heard senior leaders say numerous times and I truly believed it. I still held members accountable and pressed them to perform better when I needed to. However, my criticism of them didn't overshadow my actions to connect with them. I never wanted anyone who worked for me to feel that they were not valued or that they didn't matter.

I think an important back-story to all of this is a phrase I heard often while I was in the Army. "Your problems are your business, handle your business." In September of 1992 I entered Army boot camp. After boot camp and Advanced Individual Training at Ft. Leonard Wood, MO, I was assigned as a light wheel vehicle mechanic to Headquarters Troop 3rd Squadron 2nd Armored Calvary Regiment at Ft. Polk in Louisiana. This unit was filled with seasoned and hardened soldiers. For example, my First Sergeant served three tours in the infantry in Vietnam. I knew if we were on a battlefield, he would have my back without question. He would be willing to give his life for mine and in return I would give mine for his. The problem was that we were not at war, we were not on a battlefield. We were in a motor pool fixing trucks. If there was a problem outside of work, something as simple as a dental appointment, the standard phrase applied…"Your problems are your business, handle your business." No one wanted to hear about my problems or what I was feeling, especially if it wasn't directly work related. If my command didn't want to hear about my dental cavity, then I knew they did not want to hear about me being 18 years old and home sick. They definitely didn't want to hear about car loan with 21% interest for a used Kawasaki 600R. I actually bought that motorcycle because I didn't know what 21% interest actually was. It didn't take long for me to I decide that the Army wasn't for me so after my first enlistment I decided to not reenlist. I joined the Army National Guard but quickly found it wasn't much different and my stay there was also short. Please understand that the friends I made and memories I have with them, I would never trade. I still root for the Army during the Army-Navy games. I just never felt important. There were even a few times I hid behind a locker skipping morning formation and no one even noticed.

In January of 2004, I joined the US Coast Guard and my first unit was Station Calumet Harbor in Chicago, IL. The very first day I reported, I knocked on the Commanding Officer's door, saluted, and said *"Seaman Kuzik reporting as ordered."* The Commanding Officer, who was a Warrant Officer, said

"What the hell is wrong with you? We don't salute indoors." In the Army we did. He proceeded to tell me his expectations and as I was about to leave, I stood up and said, *"Thank you Chief, I look forward to working in the Coast Guard!"* He stood up, slapped the desk, and yelled at me that he wasn't a Chief. He was a Warrant Officer and he told me to tread lightly. I was completely knocked back on my heals, I was lost, and wondering what did I do? Warrant Officers were called "Chief" in the Army as a sign of respect. Well, right then I knew the Coast Guard was going to just as awesome as the Army was. There was zero wind in my sails, all the excitement and eagerness I had was gone. After a year, there was a change of command. Our new Commanding Officer was Steve Cabral. Mr. Cabral changed everything for me. He brought me back to life. Not just me, the entire unit. At the time I didn't even know how he did it, it just naturally happened. It wasn't until I started leading people myself that I actually understood what Mr. Cabral did. He became the leader I modeled myself after.

Later in my career, I was assigned as the Executive Petty Officer of a small boat station in upper Michigan, I had the habit of randomly pulling one a junior member in my office, and just talking to them. Just as Mr. Cabral had done to me. Plenty of times I had to reassure them that they weren't in trouble, or that they weren't going to get in trouble because the work wasn't getting done. Again, just as Mr. Cabral had done for me and many others. I let them know that everything was fine, going to continue to be fine, and that I simply wanted to talk. I used this time to get to know the members beyond the workplace. I asked about their families, where they grew up, what brought them into the Coast Guard and more. These talks became known around the station as the "Boondoggle with the Chief" and I could over hear the crew joke about it. I kept doing it though because I felt like it was the right thing to do to stay in tune to what was happening with my people and at my unit. I was intentional about earning their trust by asking questions about themselves, their families, and even their pets. I wasn't pretending to care, I genuinely cared. Just as Mr. Cabral did when he asked me about my dog that had to be put down. I wanted to know if they had a cavity. I wanted to know if they were about to sign a car loan with 21% interest. I wanted them to know that their problems were not just their business. It was our problem and our business, and we could get through it together. I had to make sure they knew I was there for them.

One Saturday evening, during the middle of a snowstorm, I received a phone call about one of our members that was in need. This member was facing some challenges in his personal life, began to drink some alcoholic beverages, and started to have some suicidal thoughts. Without hesitation, I jumped into my vehicle and headed 21 miles towards this member's home in the middle of a blizzard. That evening, another unit member and I stayed with this member overnight. We talked, told stories, laughed, and cried.... together. We let him know that we truly cared for him. When the alcohol started to wear off, his outlook on life started to brighten. His time spent in that very dark mental place had ended and I went home.

One life saved and the Search and Rescue alarm didn't sound.

Word quickly traveled around the station that I truly did care about the members. They began to realize that I wasn't the type of leader that only showed care for someone at work. I was the type that showed care for them at any time and anywhere. And they started to come to me more often. As time went on, I knew I was having an impact because my members had no problem coming to me when they had problems. That was exactly the type of leader I had once before, what I wanted to be, and was intent on becoming.

One day one of my crewmembers knock on my office door and asked if she could come in and close door. "Of course!" I quickly responded. She entered my office, closed the door, sat down in a chair, and broke down. She just started to cry. I jumped out of my chair and knelt beside her. I had no clue what she was going through but I was prepared to hear about something truly bad. I asked her "Are you okay?" She said she wasn't okay and then I asked, "Is there anything I can do?" Through the sobbing and the tears, she was barely able to utter her simply request. "I need a hug!" I will be honest in saying that I was a little surprised with the request. I was prepared to call some mental health professional, prepared to start a sexual harassment investigation, or help her get home to celebrate the life of a close family member. She didn't need any of that. She was just homesick, truly missed her family, and I reminded her of her father. She just needed to cry it out and receive a fatherly embrace. And that is what she got. Never imagined that I would be in that position, but I'm glad I was and I would do it again if the situation called for it.

She eventually stopped crying shortly after the hug, wiped away her tears, and went back to work with a rejuvenated spirit. Sometimes people just need to cry it out and that is okay. I would rather have my members come to me to cry versus having them cry because of me. Several days later I saw her working at the station one afternoon and asked how she was doing. She said she was doing a lot better and told me something that I didn't realize. She had also went to that same very dark mental space and had some suicidal thoughts. She thanked me for being there for her when she truly needed it. I thanked her for trusting me and giving me the opportunity to be there when she needed it.

Two lives saved and the Search and Rescue alarm didn't sound.

Different day but the same knock on my office door. This time it was one of my male crewmembers requesting to come in and have a closed-door discussion. I gave the same response "Of course!" Once again, I prepared myself for anything that was about to happen next. Once again, I was caught off-guard. This member was having problems with low testosterone, and it was causing some stress at home. He was embarrassed and wasn't quite sure what to do. How did I respond? With Transparency and honesty is how I responded. I told him about the time I also had low testosterone and it was nothing to be embarrassed about. It happens to a lot of men rather they want to admit it or not. I told him to make an appointment with medical just like I had done and get some help. He took my recommendation and is now about to become a father for the first time. Another situation that I didn't expect to be in but glad that I was.

Another storm arose and another knock on my door.

Another crewmember needed to have a closed-door discussion and like every time before, I was ready and willing. "What is going on? Is everything okay?" I asked prepared for any response I would hear next. I thought I was prepared but I wasn't. He looked at me and said "No, I'm not alright and I won't be alright until you tell me what is going on with you! You haven't been the same lately and everyone can see that. We need our old Chief back." At that point, my office door flew open, and a couple of other station members came in. "We aren't leaving until you tell us what is going on" one of them said. I was shocked and speechless. I thought I was doing a good job covering up the storm that I was facing. I was wrong.

A few weeks prior to that moment I received some devastating news during a medical appointment. I was told that I may have Amyotrophic Lateral Sclerosis (ALS) disease and the news crushed me. I knew there wasn't a cure.

I knew the rapid decline of my functional abilities would be tough on my family. I knew I didn't want them to go through that. I didn't want to be a burden. I had decided that the easy road was to spare them that long drawn-out pain of watching me wither away. I went to that very dark mental place and decided to take my own life.

I made the decision and thought I could just "act normal" while I got some things done before I executed my plan. I couldn't and my crew noticed a change. They could tell that something was off and that I wasn't checking on them as much. I went from being loud and talkative to reserved and secluded. They noticed and there they were, in my office staring at me, and demanding answers.

I couldn't hide it any longer. I had to open up to them, just like I had expected them to open up to me. I told them about my diagnosis of ALS. They asked if I was thinking of committing suicide and gave them an honest response. They told me that they couldn't let that happen. They showed me that they cared too much about me to sit on the sideline while I executed my fatal plan. They were intrusive with me, and it saved the life closest to me…my own. It was in that moment that I realized how much of an impact that I had on my crewmembers. I discovered how much they truly cared about me. They got me to rethink the impacts on my family if I had carried out my plan. They got me to change my plan. Thankfully months later I was told I in-fact did not have ALS and there was simple cure to my ailment. I don't believe I would have ever gotten the chance to know that if my crew members, my shipmates weren't comfortable enough to have such a bold and difficult conversation with me.

Three lives saved and the Search and Rescue alarm didn't sound.

I never would have considered myself an intrusive leader. If I would have been asked to define intrusive leadership in the past, I would have said something like micromanaging or associated the term with something negative. I recently had a discussion with Marcus Canady and read an article he wrote on intrusive leadership. I quickly realized that I could not have been more wrong. I was doing what came naturally to me and what I felt was right. I agree with Marcus when he says, "What comes naturally for some does not come naturally for everyone." Having been properly introduced to the term, I would now characterize myself as an intrusive leader and I think everyone should become one. The alarm is sounding in our organization and other organizations for this style of leadership to be the standard and the expectation. Unfortunately, Marcus considers me the exception.

What I did wasn't special or extraordinary and I felt a little embarrassed when he said that. All I did was be intentional in showing my members that I cared for them. I'm not writing to shed light on me and my actions. I'm writing this to shed light on this intrusive leadership style and tell everyone to take the same actions.

Recently someone very close to me reported to their first Coast Guard cutter. They were completely lost having only been to small boat stations in the past. They tried to discuss some situations which made them feel uncomfortable and the response they were given by their immediate supervisor had me completely disturbed. Disturbed to the point that I feel it was vital to shed light on toxic leaders who are still leading in 2023. The response was "I don't give a (expletive), it's my job to make you hate this place." I instructed the member to go to the Chief, but they felt so uncomfortable that they couldn't. Well, I did without their knowledge and hopefully it was resolved. Our members deserve to feel valued and supported as a whole person. The organization must give them that. They must develop intrusive leaders.

Thank you Chief Kuzik for your humility, transparency, compassion, and care. Thank you for your Intrusive Leadership!

Now, I am not saying that Intrusive Leadership is the perfect method that is going to eliminate suicide. I am not saying that Intrusive Leadership can replace the need for hotlines or professional mental health providers. Not even close. What I am saying is that Intrusive Leadership can create the type of relationship with people that can enhance a leader's ability to recognize when a member is in need. It can increase the likelihood that a member would indicate to someone that they need help.

It might even prevent a person from even getting to the point where they think they are so alone. Naturally, I don't k It is impossible to understand everyone's relationship with their parents, siblings, or friends. What I try to do is make sure every one of them knows that they at least have one person, then cares if they show up on Monday. There is someone who cares if they are happy or not. Someone who knows that their life matters. If there is only one person then that one person is going to be me, and I will be intrusive every time to make sure they understand that.

PART VI

Frequently Asked Questions

At the time that I am writing this 2nd Edition, I have conducted over 100 training seminars all over the country. I have conducted these trainings at military units, Federal agencies, Fortune 500 corporations, universities, and leadership conferences. Following these seminars, there have been some questions asked by multiple attendees. I wanted to take some time to answer these questions now.

Question 1: The word intrusive still has a negative connotation so why do you continue to use the word when describing this much needed leadership style?

I continue to use the word intrusive for three reasons. The first reason is that term accurately describes the actions leaders need to take when one of their people are facing a challenge and choosing to suffer in silence. Leaders need to be a little invasive, a little meddlesome, and intrude with care in the hopes of getting their people to open-up. How you do it is important, but the need is always present. There are simply times that leaders will need to be intrusive to determine what is going on within their unit, and individually with their people.

The second reason is these actions that I am describing are the same type of actions your best friend or close family would take if they noticed that you were going through something. They would ask you *"How are you really doing? Is everything really okay?"* We don't have a negative reaction when they do it because there is a level of trust that is already established. You know your best friend or close family member is coming from a place of care. This level of trust is what leaders need to establish first, so there isn't a negative reaction to their intrusiveness.

The third reason is that the term intrusive makes some people feel uncomfortable, and that is okay. Leadership is often uncomfortable. Any person that looks to stay comfortable while leading will more than likely be just managing. Leaders must get comfortable with being uncomfortable. They must be willing to have difficult conversations, hear honest feedback from their people, and discover unfortunate realities.

Almost every time I have addressed this question during my seminars, I ask the person who still doesn't like the term if they agree with the actions that I am telling leaders to take. I even asked a guy who was shouting out his disproval of the term from the back of a conference venue in front of thousands of people. He, along with every other person, said that they agreed with everything that I am saying about what leaders should be doing to connect with their people. Why is it that we agree on so much yet spend all of our time discussing where we don't agree? Or what we have a problem with? I have never scheduled someone for Intrusive Leadership counseling or told someone during counseling that I was about to get intrusive with them. I just did the actions. So, it is okay if you can never fully get comfortable with the term, just please do the actions.

Question #2: People don't want you in their business every day, so how do you avoid being too intrusive?

I agree that people don't want you asking them about their home life every single day. There is a risk of being too intrusive, so a balance needs to be achieved. You need to be intrusive but not crazy intrusive. When I say that Intrusive Leadership is the everyday leadership style for diversity, equity, and inclusion or that it should be the foundation of your leadership, I am referring to a mindset in how you approach leadership. Every day you need to be cognizant of how people are doing. Every day you need to be mindful of opportunities that may arise to connect with your people. Every day you need to know that today could be the day that you impact someone for a lifetime. You simply never know when someone is going to come into work after hearing some devasting news the night before. Or hearing some amazing news. Additionally, you don't know when someone is going to share a career goal or passion they want to pursue. Leaders need to actively listen every day during every interaction to pick up on certain things. You don't need to be intrusive with everyone every day, but you can still be an intrusive leader every day.

Question #3: How do you deal with the emotional toll that comes from being the leader that people bring their problems?

I will be honest and transparent with you in that I don't have a technique or method that eliminates the emotional toll that potentially comes from being an intrusive leader. When people share heart-wrenching stories about things that are happening in their personal lives or toxic work environments, you can't help being emotionally impacted yourself. You feel their pain. You feel their frustration. You are hurt when you see them hurting. This emotional toll is real, and it can't be ignored.

Part of the emotional intelligence that a leader must have, is understanding their own emotions. You have to be mindful of your own mental health when you are offering support. There is a reason why a lot of psychiatrists have psychiatrists. Although I am very aware of this emotional toll and try not to carry these burdens myself, it gets tough at times. When those times get tough, I confide in trusted mentors, being careful not to violate the trust of a member that confided in me. I take mental health breaks and go for runs to clear my head. In the end, I remind myself that I can't take on someone else's problems and emotions as my own to the point that I can't perform. Or to the point that it impacts me at home. I just pray on it, and let it go.

Question #4: How do you create personal relationships with each employee if you lead a large number of people?

There comes a point as a senior leader when leading people becomes difficult. When I was the Commanding Officer of Coast Guard Air Station Houston, over seventy people were assigned. I did not know each person intimately. That was not my job to do that. My job was to establish a culture of intrusive leadership at the unit. If you remember from the previous section, my passion for Intrusive Leadership began during this assignment in Houston so I will admit that I was not specifically focused on this task. My leadership philosophy had tenants of Intrusive Leadership that I used during this assignment. I just wasn't telling my middle managers to be intrusive leaders. I set the tone that my expectation was for the middle managers to lead in an intrusive manner, and I held them accountable for that. When there were performance issues with unit members, we made sure that the real issue wasn't something at home. We were intentional about earning their trust and making sure every member felt valued. Knowing what I know now, my leadership decisions could have been better.

Senior leaders are responsible for making sure the middle managers in the organization are leading in an intrusive manner. The best way to do that is for senior leaders to be intrusive with the middle managers and hold them accountable for being intrusive with the people they are directly leading. This culture should be clearly stated in an organizational leadership philosophy and understood by every member. Most importantly, there should be specific training in intrusive leadership, so people are clear on what it is and how to do it.

Question #5: How do you avoid getting a negative reaction to Intrusive Leadership if a person decides to remain private and not share personal matters with leadership?

I don't think that the risk of receiving a negative reaction to Intrusive Leadership can ever be eliminated, but it can be reduced. It can be reduced by paying attention to how and where you as a leader try to build a personal connection with someone. Connecting with a private person is going to take careful calculation and time. It isn't going to happen after the first interaction. A strong connection might not ever happen. There are some people who simply don't trust people in leadership positions enough to talk about their personal lives. Whether it is because of a previous supervisor or something that happened in their childhood. There are people who are going to only let you lead the employee part of them. What is important to make clear is if an employee opens up about a personal matter, as their supervisor or leader, you are there to offer advice or a listening ear.

It is important to understand that there is a risk in not making the effort to connect with a person who remains private or anyone for that matter. We already know that there are people who will choose to suffer in silence. Allowing them to do so could have dire consequences. I would much rather receive a negative reaction trying to connect with someone, versus receiving some negative news about them and regretting never trying. There is a risk in trying to connect and there is a risk in not trying. I chose the risk of trying each and every time.

Question #6: How do you become an Intrusive Leader, if you are an introvert?

Introverts are not going to like this response. I don't know how to make this easy for people who simply would like to avoid making personal connections with people. : Introverts simply need to identify ways to confront their fears and practice until their actions are natural. If someone wants to maximize their impact as a leader, then my opinion is they must find a way to connect with people. The Geezus technique is a great way to do that. At the end of the day, the necessity of building personal connections with employees remains a factor whether the leader is an introvert or extrovert. Introverts simply need to choose to work a little harder in accomplishing this goal.

They might have to figure out what things make them comfortable when engaging with people, or what might motivate them to make the initial effort. Introverts simply just have to find a way to make it happen if they want to be that leader.

If they can't get past being an introvert, then their only option is to find someone who can fill that role for them on the team. They might not become that leader themselves, but they can still make sure that their folks have someone that they can go to. They can simply confide in the extrovert within the unit or another staff member with better communications skills. That chosen person can simply report back to the supervisor if there is anything significant to report. The introvert might not be the intrusive leader but that doesn't necessarily mean the people can't get Intrusive Leadership.

PART VII

Intrusive Leadership in Civilian Career Fields

So, at this point I have explained my journey towards becoming an intrusive leadership champion, I have defined the leadership style, provided characteristics of an intrusive leader, given examples of Intrusive Leadership in action, and explained how impactful Intrusive Leadership can be. A large majority of my examples and stories highlight aspects of Intrusive Leadership in a military environment. That is mainly due to the fact that I have been in the military for over 22 years and my professional network is primarily made up of military professionals. Some would say that military leadership is applicable in any environment or organization. Leading people is leading people no matter the situation. There is some validity to this notion, but I don't want to end this book without a section that is focused on discussing how effective or how impactful Intrusive Leadership can be in other environments. I don't want you to finish this book and still ask yourself "Is this leadership style applicable where I work?" Let's answer that question now.

Law Enforcement

While I was going through my own struggles trying to cope with the racial tension in 2020 following a few tragic situations involving local law enforcement officials, the primary focus leading my unit was to make sure we could effectively conduct our mission and do it safely. I didn't want my aviation crews to be debating social justice issues in the break room and have that tension in place as they flew offshore in the middle of the night to save someone's life. I needed them to focus, and I needed them to trust each other as they worked seamlessly together. I felt it was important to have these difficult yet important discussions at work in order to increase workplace unity. How did that translate over to the law enforcement community? A lot of these debates and discussions involved law enforcement and how they conducted their duties. There was a direct correlation between the racial tension and their mission. They must have taken a moment to have these discussions as a department or precinct right?

During my research I talked to an African American woman who worked as a police officer in Central Texas. She explained to me how frustrated and upset she was after seeing the video of George Floyd's murder. She had a lot of questions. She had a lot of emotions stirring up inside of her. She looked at her uniform and she looked at others wearing the same uniform and had concerns. Every day she showed up to work, her life depended on the other men and women that she worked with. She had questions about how they felt about what was happening with all the protests in society over police brutality. She wondered if her life mattered to them. She never got those answers. She never had those emotions addressed.

To my surprise, she explained that there was never an attempt by her supervisors or leadership team to reach out to her or any of the other police officers and ask how they were doing personally. They discussed the job. They discussed how they were going to handle the protests in their jurisdiction. But they never intentionally reached out to find out how was she doing, while at the same time processing everything that was occurring. I asked her what impact Intrusive Leadership could have had in that moment. I asked her how it would have made her feel if someone reached out to her and said, "I'm just checking in on you to see how you were doing personally?

I know it must be difficult being a law enforcement officer, especially an African American law enforcement officer right now. I just wanted to ask you how you were dealing with all of this and see if you wanted to talk." She said that would have made all the difference in the world. She needed to be seen in a specific way by her supervisors and by her coworkers. She needed someone to show that she was valued and cared for as a person and not someone who just had a job to do. She needed Intrusive Leadership and unfortunately, she didn't get it.

I also talked to Commander Randy Griffin, who currently works for the D.C. Metropolitan Police Department, about how his department handled this situation.36 He explained to me that at the time he was leading a precinct and there was a meeting amongst the leadership team in the department. They watched the video of the George Floyd murder together in its entirety. There wasn't much discussion, but the Chief of Police at the time made it clear that what they saw was against department policy and accountability for the police officers involved was necessary. It was his opinion that what they witnessed in the video was murder and he made that clear. Commander Griffin explained that during the meeting he remained silent because he was struggling to control his emotions and he felt it was best not to say anything. Others spoke about their thoughts and the door remained open to engage on this topic. This meeting was with the department leadership team, and they were given the choice to have this discussion at the precinct level.

There wasn't an order or mandate to have the discussion, but they were given the choice and the backing of the police chief if they decided to. Commander Griffin decided to have this discussion with the men and women who worked for him at his precinct. He felt like it was important for them to hear from him and more importantly that they hear from each other. He processed his own emotions and then he authentically stood in front of them and talked. The discussion was open and honest. Members of his precinct expressed their feelings as law enforcement officers and as concerned citizens. The formal discussion lasted for a short time period, but Commander Griffin witnessed the conversation and the support for one another taking place long after the discussion was over. Through this difficult time, they had a difficult discussion, and they came together. This happened at one precinct, but it didn't happen at all of them.

Commander Griffin and I talked about Intrusive Leadership in a broader sense and his thoughts on the leadership style in regards to his experiences as a law enforcement officer. While we had this discussion, he recalled two situations. The first situation took place fairly early in his career. He was up for his first promotion at the same time that he was going through some challenges away from the job. Commander Griffin had a child with a woman that he was romantically involved with. When that relationship ended, the woman started making allegations of domestic abuse. Department policy required there to be an investigation completed anytime an accusation has been made. Commander Griffin knew that the accusations were false, and he wasn't concerned about the results of the investigation, but he was still concerned about the impact on his reputation and ability to promote. He wondered if the department would elevate someone who was going through a difficult domestic situation that could be considered as a distraction to the department.[48]

---

48. Commander Randy Griffin, Interview with DC Metro Police Commander, September 18, 2022.

He decided to talk to the Assistant Chief and the head of Internal Affairs about it, so he scheduled a meeting. He explained his concern and his supervisor provided a "textbook" response. His supervisor simply told him that he shouldn't worry about his promotion possibilities and continue to perform well. Commander Griffin remembered leaving the office still concerned and questioning the perceived lack of compassion and care in which his supervisor addressed his concerns. He left with more concerns than he had when he arrived for the meeting.

A few months passed and Commander Griffin was working a homicide scene in a local neighborhood. When the Assistant Chief arrived, he began to provide him with the routine situational brief to inform him of details involving victims, potential suspects, witnesses, and evidence collection efforts. As he was executing his duties as a police officer, his supervisor interrupted him and wanted a status update on the person. "Randy, I can get all of that information later. How are you doing? I remember you came to my office concerned about your domestic situation and your promotion a while ago. How are you doing?" Time stopped momentarily for Randy. He thought that his concerns and what he was going through wasn't visible to the supervisor. It was at that very moment that his supervisor let him know that his thoughts were wrong. The Assistant Chief, who was responsible for nearly 3800 sworn officers, did see him beyond the uniform. He saw him as a person dealing with a tough personal situation. Right there, in that moment doing his job, Commander Griffin felt supported and valued as a human being. That feeling was impactful. That feeling energized him to work harder for the supervisor, for the department, and for his fellow police officers. He wanted them to have that same feeling, so he modified his own leadership style. He was positively impacted by Intrusive Leadership, and he remembers that impact today.

Years later, Commander Griffin had moved up the ranks and became a supervisor himself. He was facing a situation at his precinct where there was some frustration in the local community. A civilian was being pursued by police officers for suspicious activity and that civilian was fatally struck by a vehicle when he ran across a busy street. The local community questioned if the police officers had a valid reason for suspecting a crime was being committed in the first place. As community frustration increased, demonstrations in the street began.

This was during a time where there was already some heightened tension in the country between law enforcement and minority communities. There was a planned protest outside the precinct building one evening and the safety of the protesters and police officers were of grave concern. If this situation escalated, then additional loss of life and property damage could easily be the result. Commander Griffin knew this reality. His supervisors knew this reality as well. They reached out to him to see if he needed anything, to see if he was prepared, to see if he was in control of the situation. He did need something. He did need some support.

At that moment, his focus was split in two directions. One direction was towards fulfilling his duties at the job, the other direction was towards fulfilling his role as a son. About twelve hours after the planned protest, Commander Griffin was going to lay his father to rest. I never met his father but as I listened to him describe his father, I wished that I did. Commander Griffin's father was clearly an integral part of everything that he has accomplished in his life and losing him hit deep. He lost his best friend and role model.

What also hit deep was the fact that nobody that he worked with on a daily basis reached out to check on him. His immediate supervisor never asked how he was coping. None of the assistant police chiefs asked if he was okay, and that hurt. The only person that showed that level of care was the same Assistant Chief that exhibited Intrusive Leadership with Commander Griffin before. One day after a morning crime briefing the Assistant Chief, now serving as the Chief of Police, called him into his office. The Chief of Police was briefed on the passing of Commander Griffin's father and made it a point to discuss his own emotions when he lost his father. He showed care, empathy, and also encouraged him to continue making his father proud. That heartfelt conversation was exactly what Commander Griffin needed and he got it from the Chief of Police.

He remembers the pain he felt when thinking that the people closest to him, his co-workers and immediate supervisors didn't truly care about him. He also remembers his desire to sacrifice and put forth his best effort took a hit. What this story tells me is that the Chief of Police was an intrusive leader, but he failed to instill Intrusive Leadership in the people he supervised. How many other police officers needed Intrusive Leadership like Commander Griffin and never got it?

When I asked Commander Griffin if Intrusive Leadership could be significant in leading all police officers in any precinct, his answer was simple. He said "Yes."

Coaching

Athletic coaches are often evaluated by wins and losses. The best coaches in any sport have resumes that are filled with championship titles. To get those titles they have to have great players that give maximum effort and make a lot of sacrifices. They have to teach, develop, and motivate. They have to get them to believe in themselves, believe in each other, and believe in that championship goal. I can't tell you exactly how they do it. If I could, then I would definitely be doing it myself. What I can tell you is that there is a constant theme when I hear players talk about their favorite coaches.

When I listened to NBA hall of fame player Allen Iverson talk about John Thompson Sr. who coached him at Georgetown University. Or, when I listened to Marcus Paige talk about his coach at the University of North Carolina, Roy Williams. There was something special about the way these men talked about their coach. They didn't describe how the coach improved their shooting abilities or how creative they were designing plays. What they discussed was how their coach impacted them beyond the basketball court. They talked about how their coach made them better men. They talked about them like you would talk about a father. You can not have that type of impact or impression on someone and focus solely on the Xs and Os of the game. To have that sort of impact, a coach must inspire and support more than the athlete. These coaches as well as all the best coaches embody the principles of Intrusive Leadership.

While I was surfing the internet one afternoon, I came across a video about a female gymnast and her college coach. This video highlighted this special relationship between a coach and an athlete and provided a perfect example of Intrusive Leadership. It was a Goalcast video in 2021 on Katelyn Ohashi. Katelyn Ohashi was considered a child prodigy in the gymnastics world at a very young age. She was known in gymnastic circles nationwide and considered as an Olympic hopeful. Recognizing her talent and potential, her parents started to invest in it. The training started to get harder and harder.

The time spent in the gym practicing started to get longer and longer. The pressure to perform started to get bigger and bigger. Somewhere along the way, the joy of doing gymnastics started to fade. Katelyn began to enjoy competing and being a part of the gymnastic world less and less. Being an athlete stopped making her happy and her performance as an athlete started to be impacted. She started to struggle in the gym, and she started to struggle outside the gym. Life was no longer fun.

Then she met Coach Valorie Kondos Field at the University of California Los Angeles. Their interactions began with conversations and Katelyn described these conversations as having nothing to do with gymnastics. Coach Field made it a point to only talk about gymnastics in the gym. Outside of the gym, the conversations were only about Katelyn and her life. The more and more they talked about life the more Katelyn started to open up. She started to trust her coach. Coach Field started to provide Katelyn with a type of care and attention that her heart and soul needed.

As time went on, the odd and uncomfortable conversations with her coach became necessary and valued. She started enjoying being at the gym because she began to love being around Coach Field. When the joy in her life came back, the joy doing gymnastics came back as well. Towards the end of the video you can see Katelyn doing an amazing floor routine. Amongst the flips and twists, what I saw was someone doing something that brought her pure joy and ended with a great smile on her face. It was one of the best floor routines I have watched and for me, I didn't have to wait to see the judges scores. Coach Field might not consider herself as an intrusive leader, she might not even know what it is. She is one though, and her Intrusive Leadership as a coach made her one of the best. In the video she states that the first thing she tried to do with Katelyn is build trust and let her know she cared for Katelyn as a person, not just the athlete. The family atmosphere that Coach Field instilled at UCLA made her a coaching legend. She won seven national titles during her career, but she transformed the lives of so many athletes along the way.[49]

---

49. Katelyn Ohasni: How Body Shaming Drove World's Best Gymnast to Quit, vol. Goalcast YouTube Channel, 2021, https://www.youtube.com/watch?v=G7sLC_XGkOM.

Teaching

When I was at the US Coast Guard Academy, I faced a lot struggles… especially my freshman year. I was a young kid from Savannah, GA who didn't like cold weather, now living in Connecticut and dealing with a lot of snow. Even though my father was in the military, I never felt like I was in the military. At the Academy, I felt like I was in the military for sure. The food was good, but it still wasn't home cooking. Like many college freshmen, there were a number of adjustments that I had to simply get accustomed to. There were some long days and the care packages my mother would send helped a little. They would have probably helped a lot if they were as big and elaborate as the care packages my sister received at Florida A&M University, but that emotional struggle is for another book that I might write one day (#NeverForget).

One of the other struggles was chemistry. Now hate is a strong word, and I really don't like using the word, but I just don't know how else I can describe the feeling that I had about that subject. I hated reading about it, hearing about it, and studying it. But I loved being around Lieutenant Eric Kowack who taught it. Lieutenant Kowack was one of the many active-duty military officers who was assigned to be a professor at the Academy and fortunately for my class he taught chemistry. He was fun to be around, and he somehow made being in chemistry bearable. We connected and I felt like he cared about me. I started paying attention more in class. I started attending his after-hours tutoring sessions. I started doing better. Clearly I wasn't the only person in my class that felt that way because Lieutenant Kowack was voted as our Class Advisor.

I remember wanting to do better to avoid disappointing him. I also remember us talking about the hottest new music album that just came out, Harlem World by Mase. I was fortunate because Lieutenant Kowack also became my high jump coach once basketball season was over. That allowed me to spend even more time with him, have more conversations, and crack more jokes. I remember it like it was yesterday.

Coach Kowack showed up to practice with a little extra pep in his step one afternoon. He was eager to tell me that he was just selected to teach the advance chemistry course the following school year. His ability to connect with students enabled him to be a better teacher and that earned him some worthy recognition. He was awarded the opportunity to teach the best students in the class. It was a good thing, but it definitely didn't seem that way to me. "Coach, those kids don't need you! They are going to be fine with anybody teaching them."

The other kids in the regular chemistry class need you, they are the ones that actually need the best teacher." I continued on some monologue about the best resources and best opportunities are always going to the best students or the well-funded schools, leaving those in need behind. I will admit that I was probably being overly dramatic and there was no need to have a Justice Thurgood Marshall moment. Coach Kowack's promotion was not another example of systematic racism. We were not at the Supreme Court of the United States. We were standing in the indoor track facility in Billard Hall at the US Coast Guard Academy. I don't think Coach Kowack was expecting that reaction or response. The surprised look on his face was clearly visible, but he heard me and he listened. His response came a few days later when he told me that he told the Head of the Science Department that he will teach advanced chemistry as long as he could still teach a couple of classes of regular chemistry. That was the spring of 1997, and twenty-five years later I still remember how that made me feel. Twenty-five years later, I still look forward to telling him about what is happening with me and hearing what is happening in his life. Twenty-five years later and I am still grateful that he taught me chemistry and so much more.

As a result of my 2020 article on Intrusive Leadership being used to address racial tension in the workplace, I was asked to speak at a number of units and participate in a few panel discussions. Most of these occasions were done in front of military audiences, but I did do one presentation for a group of college students in California. The discussion focused on the need for leaders to engage with people in a caring manner and be willing to have uncomfortable conversations about race. I still used a lot of military examples to highlight my points.

I honestly didn't have any other examples to draw from at that moment, and I was concerned if my message was resonating with the students. Then Patricia Crowley, one of the students listening to me talk, made a comment. She expressed that she understood how what I was describing as Intrusive Leadership was impactful and made a perfect comparison to how Intrusive Leadership was significant to her as a student. She, along with millions of college students, was struggling trying to maintain her academic performance while adhering to COVID protocols and adjusting to mandatory online learning.[50]

The adjustments were difficult. She explained that the teachers that took the time to check up on student's mental state and personal well-being had a lower class drop rate then teachers who did not. "The students didn't drop the classes of the teachers that showed they cared about us and how we were handling the pandemic," she said. Think about that. Students stayed in classes and successfully completed college courses in pursuit of a degree because a teacher used Intrusive Leadership.

I recently reconnected with a Coast Guard veteran that I served with and we discussed Intrusive Leadership. We discussed my message, and he also explained how it applied to him in his previous job as a high school teacher. He told me this story about a student of his who was routinely late to class and was absent a lot. The number of days absent was beginning to get close to the point that he wasn't going to be allowed to pass the course and continue to the next grade. He sat down with the kid one day and started to ask why? He didn't start by explaining the rules and telling him how he was going to be held accountable. He started by asking why was he late and why was he missing school so much? Because he asked why, he found out why. The student's mother had a substance abuse problem. When his mother was either not home or not in any condition to take care of his younger siblings, he and his older sister would take turns staying home. He wasn't just missing school because he didn't want to graduate, he was missing school to be the best brother he could be in the situation that he was in. Because the teacher asked why and got to know this student a little more deeply, he was able to help him. He helped him outside the classroom and that helped him and his sister inside the classroom. The student and his sister were able to graduate high school instead of being disenrolled.

---

50. Patricia Crowley, February 7, 2022.

One of my good friends that I met in 2002 while I was living in Pensacola, FL read the first edition of this book. He actually reads any leadership book that he can get his hands on which means he has definitely read a lot more leadership books than me. He reads every book on leadership he can find, which is significantly more than me. I'm working on getting better, don't you worry!

While Michael Biggs was attending Tennessee State University, he was hit in the gut with some devastating news. Both of his parents were arrested for drug trafficking and were facing some serious time in the prison system. This normal college kid with a bright smile and infectious personality changed. His grades started to slip, and he was less attentive in class. One of the better students studying civil engineering was now in danger of failing out of school. He was depressed and struggling to deal with the reality of both parents being convicted felons and his life changed forever.

In steps Dr. Mishu Farouk, who was one of Michael's class instructors, and was also his academic advisory and civil engineering department chair. Dr. Farouk noticed the sudden change in Michael's behavior and performance. One day he approached Michael after class and requested to meet with him in his office. Once they were in the office, Dr. Farouk started to inquire about what was going on with Michael.

At the time, Michael hadn't told anyone at the school what was happening with his parents. He was concerned about being judged and didn't want to respond to questions about this matter. He was simply suffering in silence. When Dr. Farouk asked him why his demeanor had changed and why his grades were starting to fall, Michael decided to open up. It has been over fifteen years since that conversation in Dr. Farouk's office, but he vividly remembers the weight being lifted off of him. Simply having someone recognize what was going on with him beyond the classroom and show genuine care meant the world to him.

After Dr. Farouk heard about Michael's situation with his parents, he came up with a plan that would allow Michael some time to deal with his situation and remain in school. Dr. Farouk advocated for Michael and gave him a second chance. Michael stepped away from school for the rest of that semester but returned at the beginning of the next. He returned in a better place mentally and his academic standing improved because of it.

Instead of becoming another college dropout statistic, Michael earned his civil engineering degree and is now enjoying a flourishing career. All of this because he was fortunate enough to have instructor that cared about him way more than as a student.[51]

Opportunities like this can also be missed. James Thach entered the U.S. Coast Guard Academy the same summer that I did in 1996. He was a smart kid from Connecticut and came from a family of service. His father worked for Sikorsky Aircraft, responsible for manufacturing the Coast Guard's MH-60 helicopters.   We were both running around, getting yelled at, and enjoying the beginning of our military journey. The summer ended and the Fall semester began. James' college academic career got off to a great start. He adjusted well to college life and enjoyed still being close to home. The Spring semester was a different story.

James' grades started to drop and going home stopped being an enjoyable getaway from military service academy life. James' father had lymphoma, and he was starting to go through a rough patch. The battle with this form of cancer had begun years before, and the battle included good times as well as rough periods. Each swing was another part of the emotional roller coaster. The spring semester was a downward section of the roller coaster, and it was impacting James' focus. Ironically, James had no problems scoring well on mid-period or final semester exams, but it was the day-to-day assignments and class participation where the problems persisted.

James met with teachers and his company officer, and they provided warnings about potential outcomes if his academic performance continued. The conversations were very transactional and only dealt with what was occurring in the classroom. There were no questions about what was happening at home. Since James would do well on the exams, the assumption was that he was just choosing not to put forth the effort in other areas of his classes. Even with that assumption, nobody asked why?

James' survived summer school in 1997 but the problems persisted and continued into the Fall semester of sophomore year. His father's battle continued, and James' battle in dealing with it also continued. Unfortunately, he was left to fight his battle alone for the most part. James was forced to leave the U.S. Coast Guard Academy after that Fall semester.

51. (Michael Biggs interview, June 11, 2024)

Fortunately for the U.S. Coast Guard, James never lost his desire to serve. He enrolled in a civilian college and eventually joined the Coast Guard reserves. He earned his degree in geophysics and ended up retiring as a Commander, faithfully serving this nation for over 25 years.

This story has a happy ending, but I can't help to think" with "but things could have been much better. James needed an Instructor to be intrusive and not allow him to face these personal battles alone. James should have been with our class on graduation day. He was smart enough and clearly able to perform militarily. He just needed someone to treat him as a son who was watching his father go through a difficult medical situation, versus solely as a student struggling in the classroom. This isn't pointing the finger at anyone. I was James' classmate, and I didn't seek to find out why my classmate was struggling either. I didn't offer any additional support. I am only telling this story in the hopes that it can help others identify the opportunity to be intrusive when someone could need it most. In the hopes that more teachers will be the difference between a struggling student finishing college versus being forced to leave college.[52]

Fortune 500 Companies

Just like members in the military, there are employees at Fortune 500 companies that are dealing with serious issues and are faced with huge challenges.[52] There are employees at marketing firms who are dealing with depression, kids with special needs, or elderly parents who are facing dementia. These employees need to feel valued and cared for just like our military members. Intrusive Leadership would positively impact them as well. I discussed how Intrusive Leadership could apply in corporate America with Colonel Isaac Taylor, US Army. When he brought his Intrusive Leadership style to a private corporation during a yearlong Public Affairs fellowship, the feedback he received was clear and positive.

---

52. (James Thach interview, August 7, 2024)

Colonel Taylor explained to me that one day he noticed that one of the employees was not at work at her usual time. He asked around the office if anyone was knew where this employee was or why she was not at the office. Nobody had an answer. Nobody had noticed her absence. He decided to give her a call. This is a normal thing in the military, so this was a sensible action by Colonel Taylor. This was not a normal thing in this particular corporate office so when the employee answered the phone, she was caught off guard.

She questioned why Colonel Taylor was calling and why he was asking about her whereabouts. When he explained that he was simply calling to make sure she was okay and that he truly cared about her well-being, her attitude changed. She stopped questioning why Colonel Taylor called and started questioning if anyone else would have called. She appreciated knowing that someone cared enough to check on her welfare and not only to get an update on a project or task.

The impact of his Intrusive Leadership style didn't stop there. By talking and engaging with employees, he discovered that there was an apprehension for scheduling medical appointments during the workday causing some employees to go without routine check-ups. There was a perception amongst the employees that taking time off from work for medical appointments was frowned upon. This was not the intended desire from upper management and a communications strategy was put in place to ensure all the employees knew that they were free to take full advantage of their employee healthcare benefits. Colonel Taylor was able to permanently improve this particular workplace environment, while on a temporary assignment because of his leadership style. These employees, touched by Intrusive Leadership, still reach out and show appreciation to the Colonel Taylor for how he impacted the organizational culture for the better.[53]

---

53. Colonel Isaac Taylor, May 14, 2022.

I have recently discovered the joys of advanced technology in the form of audiobooks. In full transparency, I have always struggled to read a book from cover to cover but audiobooks are a heavenly blessing sent directly to people just like me. Thank you Lord! One of the very first audio-books that I listened to was *"Where You Are, Is Not Who You Are"* by Ursula M. Burns.[54] A great book about an amazing woman who came from humble beginnings and ascended to the highest levels of corporate America. She was able to achieve success because of her strong work ethic, loving mother, and government sponsored educational programs for disadvantaged communities. After listening to this book, I came away with two messages. First message was no matter what current position you hold or situation you are in, that position or situation doesn't define who you really are as a person. It doesn't matter if the situation is a bad one or the position is a lofty one. You are not bounded by your situation and there is something more to you than your status. The second message of course involves Intrusive Leadership.

Ursula Burns worked for Xerox for many years. She was a phenomenal employee who was truly dedicated to the success of the company. She started out as an entry level employee and rose to a very senior level of the company. She had established a relationship with the CEO and was known by the company's Board of Directors. The company had invested in Burns and Burns was loyal to the company. She never considered leaving because she felt valued until she didn't. A change in company leadership brought a change in how she felt loyal to the company. She no longer felt valued, so she was no longer loyal. The company was struggling, and she no longer felt invested in the company's success.

She started looking to transition to a new company and the search didn't last long. She was well qualified, highly sought after, and quickly received a lucrative offer with a different company. She made her decision to accept the offer and started making her notifications to her supervisors and mentors at Xerox. Most of these notifications were easily completed but there was one that Burns dreaded. That was the notification to a previous CEO and a long-time mentor. They had a relationship that clearly went beyond the workplace

---

54. Ursula M. Burns, Where You Are Is Not Who You Are, 1st ed. (HarperCollins Publishers, 2021).

bounds. This mentor was an intrusive leader and told Burns something that I believe a lot of people in leadership position refrain from saying to a subordinate. "I need you." A simple phrase, but when said by an intrusive leader it is a game changer. Changes were made at the company and Burns remained at Xerox. She played an integral part in turning the company around and eventually became the CEO of Xerox herself. She actually became the first African American woman to be a CEO of a Fortune 500 company. Xerox was able to retain Burns and benefit from her amazing abilities and corporate knowledge because of Intrusive Leadership.

This past year I had the pleasure of being connected with John Pryor who currently works for a leadership development company. After reading the first edition and reflecting upon his personal experiences, he discussed with me how a culture of intrusive leadership can positively impact a corporation's bottom line.[55]

John previously worked for Southeastern Freight Lines in the trucking industry and there was a common phrase that the President of the company used to establish a leadership culture throughout the company. "We are going to be true to the wall," was the phrase and all the executives and employees knew what that meant. On the wall in every facility was a plaque that had the company's "Game Plan for Success," which provided a leadership model and vision for everyone to inspire to live up to. The phrase "We are going to be true to the wall," reminded every person who was making operational and personnel decisions how those decisions are to be evaluated, analyzed and executed. On the wall was the following:

---

55. John Pryor Interview, July 24, 2024)

Mission
Live Our Culture,
Provide Quality and
Create Financial Strength

Culture
Value Each Other Completely
We are in the people business
Integrity
Doing the right thing is always right
Serving
We are here to SERVE
Continuous Measurable Improvement
Anything can be made better
Belief, Alignment and Teamwork
A unified team is a competitive weapon

Quality Improvement Process
People Committed to Individualized Quality (IQ)
Train and Equip for Success
Team Structure and Participation
Formalized Process Improvement

Financial Strength
Perpetuate our company and provide long-term security for our team

SOUTHEASTERN FREIGHT LINES   **Game Plan for Success**

Following this game plan, the company made investment decisions with the strategic vitality in mind to give the company flexibility during potential economic downturns. The long-term security of the team included every employee. The company simply prepared for a rainy day and thankfully they did. That rainy day came when the country entered into an economic recession and the company's goal of being true to the wall was tested.

As the trucking industry faced the financial impacts of an economic recession, the company had a clear and stated goal. Up until that point the company had never had a layoff, and the goal was to keep that streak going. They didn't have a profit goal, instead they had a personnel goal. They determined their employees were family and they wanted them equipped to survive tough times. So, they made decisions and put processes in place to be true to the wall and achieve their personnel goal.

During this time, they kept communications open and found new ways to keep everyone employed. If a person was able to find additional employment through a family business, they allowed them time off and remained on the company's insurance plan. They requested employees use their earned vacation time and provide additional vacation time. The company spread the work hours around as evenly as possible to make sure everyone was able to earn money. They established employee-sharing agreements with other companies to provide additional work hours. They accepted contracts they knew were not profitable simply to generate revenue and keep employees working.

At the end of the recession, the company achieved their goals and they were able to retain every employee they had. The company had its lowest financial report only achieving a 0.4% profit margin, but it was considered their best performing period because of what the goal was. They expected to be in the negative, but they ended up in the positive without laying anyone off. When the economy took a positive turn they were in a perfect position to positively respond. They had a workforce that was already trained and prepared to move shipments and take on new contracts. While other companies had to take time to hire and train new employees, Southeastern Freight Lines was securing profitable contracts and was poised to have a high-earning quarter. How else did the company benefit from being true to the wall? The employees showed their appreciation. Phone calls, letters, and emails came flooding in from the families expressing how grateful they were to be working for a company that truly cared for them. Their employees had neighbors and family members working for other companies who were laid off and lost their family insurance plans. They saw the impacts of working for a company who didn't put their people first, so they showed their gratitude by their loyalty. Not only did the company benefit financially because they avoided hiring and training expenses and being in a position to quickly get back to operations. They also avoided legal expenses because they avoided lawsuits and unionization efforts that their competitors faced. When a new employee who wasn't around during the recession criticized the company, the older employees came to the company's defense. The amount of time spent on human resource issues remained low because organizational trust was high. Southeastern Freight Lines benefitted immeasurably by installing an organizational culture that put people first and intentionally made sure everyone felt valued. They benefitted from a culture of Intrusive Leadership.

I hope by now it is understood that this leadership style is about people. It is primarily focused on the relationships built and connections made with people. Industries or situations that involve leading people, Intrusive Leadership applies. It not only applies, but it greatly enhances the effectiveness of leaders and greatly improves the workplace climate for employees. It improves the learning environment for students in the classroom. People are the greatest asset of any company, and this leadership style places people first.

PART VIII

What Happens Now?

Well good people, that is it. That is my argument that intrusive leadership is the most critical leadership style in the future. I have explained my personal journey. I have provided the definition of the leadership style. Characteristics and examples have been explained. The impacts of this leadership style have been described, and evidence of how it applies in any industry or organization have been given. At this point, you have all the information that you should need to decide if you are going to lead this way. Right now, I would ask you to reflect on your own leadership style and if you are leading in this manner. Whether or not you can do it isn't the question because I believe that you can as long as you make the decision to try. That decision to become an intrusive leader and the decision to start making the effort to connect with your employees is all that is needed. On a personal level, I would like for you to ask yourself that question and be confident in your decision. After you make that decision, then you have to take action on that decision. As I mentioned before, how you employ Intrusive Leadership will depend on a number of different factors. Just figure out a way to engage with the people you lead, be authentic, and ask questions about them. Never underestimate how far you can go by simply starting with a simple question. How are you today?

At the organizational level, I would like for senior leaders in organizations around the world to look at their organizational culture and climate and determine if Intrusive Leadership is present. If it is not, then how do you begin to insert it. Senior leaders have to start messaging to their employees and the rest of the leadership team that Intrusive Leadership is the expectation going forward. Ensuring that every person who comes to work, whether they travel to the workplace or just log in at home, feels valued and supported as a person. Managers need to know that they are expected to get out of their offices and go engage with the people they supervise and get to know them beyond the job.

This leadership style needs to understood as the standard and that understanding has to come from strategic communication from the senior levels. Once the decision has been made, and the expectation communicated, then people have to be trained. Senior leaders, mid-level managers, and first-line supervisors need to go through a leadership course that includes a session on Intrusive Leadership. They need to know how to define it, recognize it, and understand the impacts of it. They need to know when to apply it and when other leadership styles could be used in tandem with it. You can't just tell someone to be an Intrusive Leader and expect them to automatically know how to do it right.

Engaging in difficult but necessary conversations does come with some risk. The risk is undoubtedly worthwhile, but the training is necessary to ensure that every leader or manager in the organization has a basic understanding of Intrusive Leadership and how to correctly do it. When the decision has been made, and the training has been completed, then people have to be held accountable for doing it. If people aren't held accountable for leading in the manner that has been communicated as the expected standard, then it will never be truly part of the organizational culture. Because this leadership style will force some people out of their comfort zone, because it requires some concerted effort, there will be some people who will choose not to do it if they know their career progression will not be impacted at all. If they can receive the highest performance evaluations, recommended for the next position of higher responsibility, or never told that they are not meeting the standard as a leader then the desired expectation will never be ingrained in the organizational culture.

Human resource managers need to analyze company policies and practices to see if they are holding supervisors accountable for managing instead of leading. Are feedback sessions and performance reviews part of the organizational system for evaluations? Are supervisors trained on how to provide mid-period performance feedback that aligns with the expectation for how they are to groom future leaders in the organization? Are they held accountable for doing them or are they given a choice? When facing a situation involving a low performing employee, are supervisors asked if they have made the effort to be intrusive and build a relationship with that employee?

Are people asked about a time that they used Intrusive Leadership when interviewing for a position? Instilling Intrusive Leadership into an organization must be done strategically and consistently. Organizations must invest in it and those that do will benefit from those investments. I recommend you start making those investments in yourself and in your organizations right now.

My Why?

One year later, my why hasn't changed so this section remains unchanged.... I'm sure that by now everyone has heard of Simon Sinek's book on the importance of finding your "why". The reason behind why you do what you do and discovering your true purpose or calling. So why has this become my why? Why has spreading the word on Intrusive Leadership become my passion? Simple…it has impacted me, and I have witnessed it impact others. I have had the great fortune to be involved in some life-saving missions that resulted in bringing someone home. I have seen families being reunited with loved ones. I have looked into tear filled eyes of someone that had just been rescued and listened to the gratitude in their voice as they told my aviation crews thank you.

That sense of fulfillment in helping others is mesmerizing and infectious. I will tell you that the sense of purpose, that sense of pride in being involved in something that is bigger than yourself, is the same whether it is saving a fisherman on the high seas or saving a market analyst in the boardroom. Having someone thank you for helping them define success for themselves, work through a difficult situation, support them as they mourn the loss of a loved one, provide them with some flexibility in their work schedules to allow them to be a more involved parent, or motivate them to follow their true passion provides that same sense of fulfillment. This will allow more time for motivation and displaying a true passion that can provide a sense of fulfillment.

If you are fortunate enough to be in position to prevent someone from taking their life, you will never forget that moment. I want everyone to feel that sensation. I want everyone to travel to work on a daily basis knowing that they are headed to a place where they are valued and supported. In addition to that, you can not impact people at the job this way and not be truly impacted personally at the core. Engaging in difficult but necessary conversations at the workplace and learning from different perspectives will change you well beyond the office. I truly feel that as you become a better supervisor or better leader, you will also become a better neighbor and a better friend. How you interact with people as a member of the neighborhood Homeowners Association or School Board will change. When you ask the cashier at the grocery store how their day is going, you might listen to the response a little harder. You never know when you might be in position to

Marcus Canady has developed, sharpened, and tested his leadership abilities throughout his military career of over twenty years. Now his passion is helping others do the same and leading our servicemembers in an impactful manner. He does not want to sell his book to military members that desire to enhance their leadership abilities. His goal is to be able to provide a copy of the book as a gift instead. In partnership with the US Coast Guard Academy Alumni Association, a fund has been established with the sole purpose of funding books for military members. If you would like to support this effort and donate to the fund, then please visit the CGAAA website below. Thank you for supporting our troops!!!

http://www.cgaalumni.org/intrusiveleadership

Support The Troops

About the Author

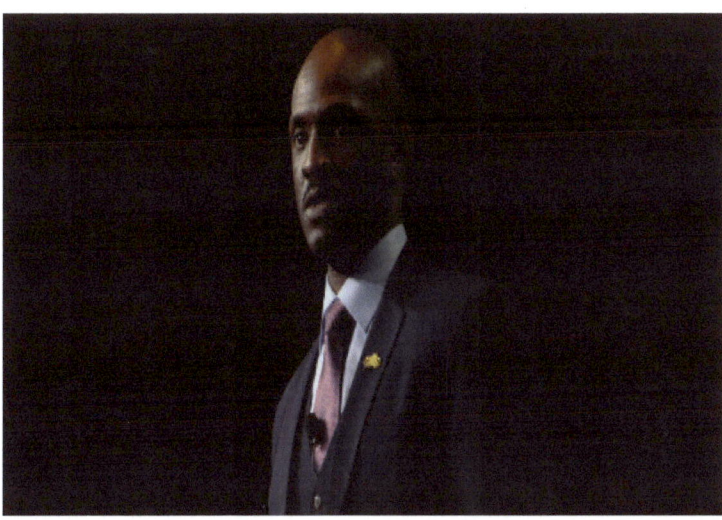

Marcus A. Canady has spent over two decades on active duty in the US military as a helicopter pilot conducting life-saving missions and law enforcement operations. In addition to hundreds of lives saved and over $250 million of illegal narcotics seized, he has spent countless number of hours leading, developing, and mentoring men and women. Having been a beneficiary of excellent leadership and mentoring throughout his life, he has turned his passion into a calling to help other leaders add to their leadership toolkit.

Marcus A. Canady graduated the US Coast Guard Academy in 2000, completed Naval Aviation Flight Training in 2003, and completed graduate school at Duquesne University, Syracuse University, and the National War College in Washington, DC. He is the son of Allen and Linda Canady, the husband of Angelisse Canady, and the father to three amazing children.

Got an idea for a book? Contact Curry Brothers Publishing, LLC. We are not satisfied until your publishing dreams come true. We specialize in all genres of books, especially religion, self-help, leadership, family history, poetry, and children's literature. There is an African Proverb that confirms, "When an elder dies, a library closes." Be careful who tells your family history. Our staff will navigate you through the entire publishing process, and we take pride in going the extra mile in meeting your publishing goals.

Got an idea for a book? Contact Curry Brothers Publishing, LLC. We are not satisfied until your publishing dreams come true. We specialize in all genres of books, especially religion, self-help, leadership, family history, poetry, and children's literature. There is an African Proverb that confirms, "When an elder dies, a library closes." Be careful who tells your family history. Our staff will navigate you through the entire publishing process, and we take pride in going the extra mile in meeting your publishing goals.

Improving the world one book at a time!

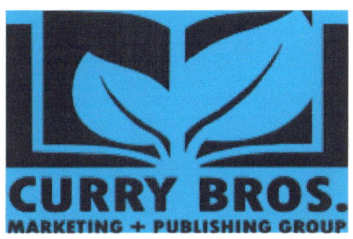

Curry Brothers Publishing, LLC PO Box 247
Haymarket, VA 20168
(719) 466-7518 & (615) 347-9124
Visit us at www.currybrotherspublishing.com

www.ingramcontent.com/pod-product-compliance
Lightning Source LLC
Chambersburg PA
CBHW040844120626
46547CB00001B/21